Lecture Notes in Mechanical Engineering

Lecture Notes in Mechanical Engineering (LNME) publishes the latest developments in Mechanical Engineering—quickly, informally and with high quality. Original research reported in proceedings and post-proceedings represents the core of LNME. Volumes published in LNME embrace all aspects, subfields and new challenges of mechanical engineering.

To submit a proposal or request further information, please contact the Springer Editor of your location:

Europe, USA, Africa: Leontina Di Cecco at Leontina.dicecco@springer.com
China: Ella Zhang at ella.zhang@springer.com
India: Priya Vyas at priya.vyas@springer.com
Rest of Asia, Australia, New Zealand: Swati Meherishi at swati.meherishi@springer.com

Topics in the series include:

- Engineering Design
- Machinery and Machine Elements
- Mechanical Structures and Stress Analysis
- Automotive Engineering
- Engine Technology
- Aerospace Technology and Astronautics
- Nanotechnology and Microengineering
- Control, Robotics, Mechatronics
- MEMS
- Theoretical and Applied Mechanics
- Dynamical Systems, Control
- Fluid Mechanics
- Engineering Thermodynamics, Heat and Mass Transfer
- Manufacturing
- Precision Engineering, Instrumentation, Measurement
- Materials Engineering
- Tribology and Surface Technology

Indexed by SCOPUS, EI Compendex, and INSPEC.

All books published in the series are evaluated by Web of Science for the Conference Proceedings Citation Index (CPCI).

To submit a proposal for a monograph, please check our Springer Tracts in Mechanical Engineering at https://link.springer.com/bookseries/11693.

Narendra Kumar · Gurraj Singh · Rajiv Trehan ·
J. Paulo Davim
Editors

Advances in Materials and Agile Manufacturing

Select Proceedings of CPIE 2023

Springer

Editors
Narendra Kumar
Department of Industrial and Production Engineering
Dr. B. R. Ambedkar National Institute of Technology
Jalandhar, India

Gurraj Singh
Department of Industrial and Production Engineering
Dr. B. R. Ambedkar National Institute of Technology
Jalandhar, India

Rajiv Trehan
Department of Industrial and Production Engineering
Dr. B. R. Ambedkar National Institute of Technology
Jalandhar, India

J. Paulo Davim
Department of Mechanical Engineering
University of Aveiro
Aveiro, Portugal

ISSN 2195-4356 ISSN 2195-4364 (electronic)
Lecture Notes in Mechanical Engineering
ISBN 978-981-99-6600-4 ISBN 978-981-99-6601-1 (eBook)
https://doi.org/10.1007/978-981-99-6601-1

This Springer imprint is published by the registered company Springer Nature Singapore Pte Ltd.
The registered company address is: 152 Beach Road, #21-01/04 Gateway East, Singapore 189721, Singapore

Paper in this product is recyclable.

Contents

About the Editors

Dr. Narendra Kumar is currently holding the position of Assistant Professor in the Industrial and Production Engineering Department at Dr. B. R. Ambedkar National Institute of Technology, Jalandhar, Punjab. He earned his Ph.D. in Mechanical Engineering from the esteemed PDPM Indian Institute of Information Technology, Design and Manufacturing Jabalpur. His research pursuits are centered around cutting-edge pellet-based additive manufacturing techniques, particularly for flexible parts. This innovative work has garnered recognition, resulting in funded research projects supported by prominent organizations such as ISRO and SERB, showcasing his commitment to pushing the boundaries of engineering knowledge. Dr. Kumar's contributions have been recognized with the prestigious IEI Young Engineers Award in the production engineering division, underlining his dedication to academic excellence and innovation in the field.

Dr. Gurraj Singh Assistant Professor at Dr. B. R. Ambedkar National Institute of Technology, Jalandhar, is an expert in Manufacturing Engineering, with a focus on sustainable manufacturing, agricultural waste management and machining. Holding a Ph.D. from the same institution, he has received the Research Excellence Award from Lovely Professional University in 2019. His dedication to advancing knowledge in manufacturing engineering, combined with practical experience and a commitment to sustainability, makes him an influential figure in the field, inspiring both students and colleagues.

Dr. Rajiv Trehan is working as Associate Professor in the Department of Industrial and Production Engineering at Dr. B. R. Ambedkar National Institute of Technology, Jalandhar. He holds a Ph.D. degree in Industrial Engineering. He has more than three years' experience in industry and 21 years' experience of teaching UG and PG students. He has guided 25 M.Tech. thesis and has five Ph.D. students. His area of interest is quality management, advanced manufacturing and has published more than 50 articles on these topics in different journals and conferences.

J. Paulo Davim is Full Professor at the University of Aveiro, Portugal. He is also distinguished as Honorary Professor in several universities/colleges/institutes in China, India and Spain. He received his Ph.D. degree in Mechanical Engineering in 1997, M.Sc. degree in Mechanical Engineering (materials and manufacturing processes) in 1991, Mechanical Engineering degree (5 years) in 1986, from the University of Porto (FEUP), the Aggregate title (Full Habilitation) from the University of Coimbra in 2005 and the D.Sc. (Higher Doctorate) from London Metropolitan University in 2013. He is Senior Chartered Engineer by the Portuguese Institution of Engineers with an MBA and Specialist titles in Engineering and Industrial Management as well as in Metrology. He is also Eur Ing by Engineers Europe FEANI-Brussels and Fellow (FIET) of IET-London. He has more than 35 years of teaching and research experience in Manufacturing, Materials, Mechanical and Industrial Engineering, with special emphasis in Machining and Tribology. He has also an interest in management, engineering education and higher education for sustainability. He has guided large numbers of postdoc, Ph.D. and master's students as well as has coordinated and participated in several financed research projects. He has received several scientific awards and honors. He has worked as an evaluator of projects for ERC-European Research Council and other international research agencies as well as an examiner of Ph.D. thesis for many universities in different countries. He is Editor-in-Chief of several international journals, Guest Editor of journals, books Editor, book Series Editor and Scientific Advisory for many international journals and conferences.

Topology Optimization of Disc Brake

Shubh Shrey, Niteesh Pawar, and S. S. Pachpore

Abstract Heat generation and friction have long been a problem in the automotive industry. The braking system is one application where these come together and benefit the vehicle's safety. Disc brakes must be able to provide sufficient friction force to stop the vehicle. To provide adequate braking force, a significant amount of heat is generated, which must be rapidly dissipated from the disc rotor or brake pad. Since the disc rotor is a rotating mass, it must be optimized because such masses have higher inertia and reduce vehicle efficiency. As a result, reducing the rotating masses will improve the overall performance of the vehicle and make it more cost-effective to operate. Topology optimization was carried out while keeping the size criteria constant and focusing solely on weight reduction as a major parameter. The obtained topology optimization geometries were further analysed and compared to the original under the same parameters, and it was discovered that the induced stresses and thermal limits were maintained with negligible changes. The optimized geometries ranged from 1.21 to 9.76% mass reduction for the five models created, for an average mass reduction of 6.23%.

Keywords Topology optimization · Disc brake · Mass optimization

1 Introduction

With the advent of electric vehicles in the automobile sector, the focus of the power generation and internal combustion engine issues have surely reduced, but the other issues which are connected to mobility continue to be prevalent in the new era. Aerodynamics, thermal dissipation and braking systems amongst others still exist with the newer vehicle systems. Regenerative braking systems are dependent heavily on the ability of the discs to provide friction and dissipate the heat readily to the environment

S. Shrey · N. Pawar · S. S. Pachpore (✉)
School of Mechanical Engineering, Dr. Vishwanath Karad MIT—World Peace University, Pune 411038, India
e-mail: swanand.pachpore@mitwpu.edu.in

N. Kumar et al. (eds.), *Advances in Materials and Agile Manufacturing*, Lecture Notes in Mechanical Engineering,
https://doi.org/10.1007/978-981-99-6601-1_1

converting the braking power into stored power for further use. Hence, disc rotors must be structurally sound as well as have high heat dissipation properties to be used effectively. The previous authors' work referred for the inspiration of this paper was studied and drawn to a conclusion that shape optimization and weight reduction are major parameters. Structural integrity and friction are the base requirements, and heat dissipation is a function of the friction and braking force produced. Umale and Varma [1] studied the deformation of the rotor along with the stress induced along with heat flux as the thermal consideration. They later selected a range of thickness of the disc rotor and tried to determine the optimum thickness for their application. Matariya and Patel [2] adopted the approach of using the calliper piston diameter along with the disc rotors. They used Ansys 18.1 and found out the equivalent elastic stress and total deformation using the material of grey cast iron and ceramic carbon composite. They optimized the application for the weight keeping the other parameters consideration to be the same. Eswaraiah and Govindu [3] utilized the parameters of total deformation, normal elastic strain, temperature, and total heat flux to determine the allowable conditions using SolidWorks for design, Hypermesh for deck preparation and Ansys for steady static-thermal analysis. Ranjith Kumar et al. [4] used the tools of Ansys, Hypermesh and Solidworks to calculate the stress fields and deformations related to the desired boundary conditions in terms such as disc pad pressure against the disc. Steady-state thermal analysis was considered for the desired thermal constraints. Thigale and Shah [5] modelled their disc rotor based on Bajaj Discover 125M disc and performed weight reduction by implementing topology optimization on Altair Inspire 9.5 and then utilized tools of Ansys 16.0 to test for constraints. They managed to reduce the weight of the disc by 26.68%. Naikwadi et al. [6] worked on the reduction of the raw material cost without affecting the function and working within the defined parameters of stress and displacement. Linear static and normal mode analysis were carried out for extreme considerations. The final design was created with a factor of safety of 1.5 [7]. In this research paper, firstly structural and thermal analysis would be done for the disc brake models. Based on the constraints applied, optimized geometry of the disc brake models would be created. These newly created optimized disc brake models would be further scrutinized in terms of their structural and thermal performances [8]. Finally, comparison would be done for both the cases, in order to prove that topological optimization does not undermine the structural and thermal performances of the disc brake.

2 Methods

2.1 Design Methodology

The 3D Modelling of Disc Brakes was done on SolidWorks 2018 software. In total, five iterations were made for the proposed study. The proposed 3D models have the following geometric specifications, as indicated in Table 1.

Table 1 Geometric specifications of disc brake models

Parameter	Description
Outer diameter (OD)	300 mm
Inner diameter (ID)	100 mm
Diameter of holes	10 mm
No. of holes	6
Thickness	5 mm

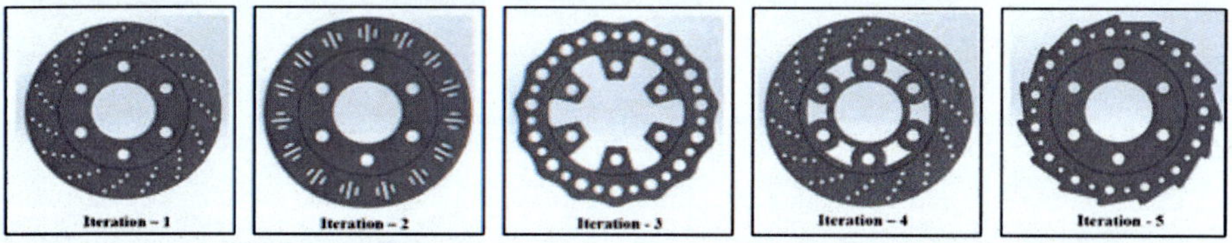

Fig. 1 3D models of different disc brake iterations

Each iteration of the disc brake model has the same specifications as stated in Table 1, with the only difference being the material removal pattern. The brakes are assumed to be made up of aluminium alloy having the following mechanical properties [9, 10].

Density: 2770 kg/m^3.
Ultimate tensile strength: 310 MPa.
Poisson's ratio: 0.34.

The different design iterations were based on various research paper resources and existing disc brake models. Figure 1 image shows the various design iterations.

2.2 *Finite-Element Method*

Finite-element analysis is a method by which the behaviour of a component can be predicted via means of various numerical and mathematical techniques. In our proposed study, static structural and steady-state thermal analysis of disc brakes was done in order to find out their structural integrity and temperature distribution under the external loads [11]. ANSYS Workbench 18.2 software was used for this purpose [12].

3D Tetra meshing was done for all the CAD models. A minimum edge length of 0.5 mm was used for creating fine mesh to obtain accurate results. In general, a total of 15,000 + elements and 30,000 + nodes were created during the meshing step, as shown in Fig. 2.

In Fig. 3, for static structural, a pressure of 500 kPa was applied on both sides of the disc brake. Also, the six innermost holes near the internal diameter were made

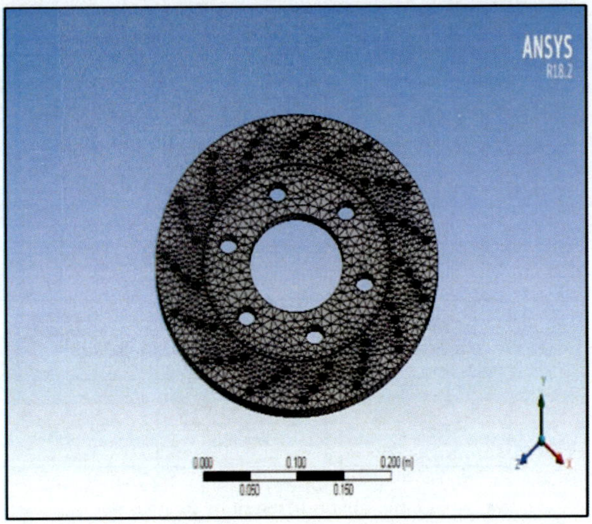

Fig. 2 Meshed disc brake geometry

as fixed support. For steady-state thermal, temperature was set at 80 °C and ambient temperature was set at 22 °C. Convective heat transfer coefficient (h) was set as 20 W/m^2 °C [9].

Topology optimization was carried out on all disc brakes following the static structural analysis. Topology optimization is a method by which, material distribution is optimized based on the applied loading conditions and design space [13]. In the proposed study, mass was considered as the response constraint with a constraint of 50% retainment, as shown in Fig. 4. The optimized geometry was then again

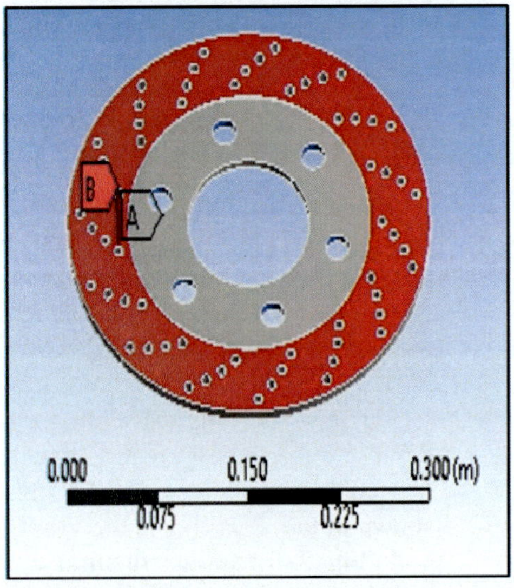

Fig. 3 Boundary conditions for static structural analysis (A: fixed support, B: pressure)

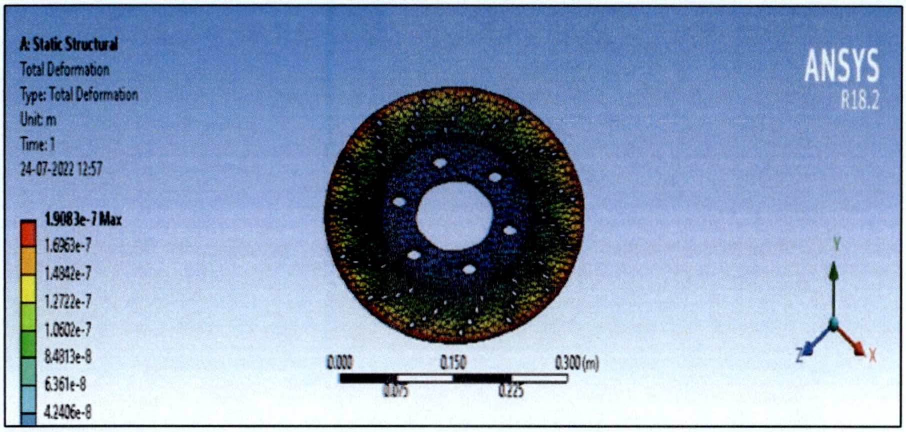

Fig. 4 von Mises stress (iteration 1)

subjected to static structural analysis to find out the effect of topology optimization on stress and deformation of the component.

For analysis setup of topological optimization, the analysis was carried out to analyze the deformation, and represented in the regions shown in Fig. 3, A was the design region and B to be the exclusion region.

3 Results and Discussions

The FEA performed subjected to the mentioned boundary conditions in preceding sections; the following results were obtained for static structural analysis before topological optimization.

In Fig. 4, for Iteration 1, maximum von Mises stress and total deformation of 0.603 MPa and 0.19 microns were observed, respectively, during the static structural Analysis.

Similarly, maximum von Mises stress and total deformation of 0.575 MPa and 0.181 microns were observed for Iteration 2 and 0.569 MPa and 0.219 microns were observed for Iteration 3, respectively, during the static structural analysis.

Table 2 summarizes the results obtained for all the iterations of the disc brake models for static structural analysis (before topology optimization).

The following results were obtained for steady-state thermal analysis before topological optimization.

In Fig. 5, for Iteration 4, maximum and minimum total heat flux was observed to be 46,137 W/m^2 and 10,253 W/m^2, respectively, while the lowest temperature observed was 74.761 °C.

Table 2 Summary of static structural analysis (before topology optimization)

Parameter	Before topology optimization		
	Mass (kg)	Total deformation (microns)	von Mises stress (MPa)
Iteration 1	0.747	0.19	0.603
Iteration 2	0.725	0.181	0.575
Iteration 3	0.494	0.219	0.569
Iteration 4	0.684	0.207	0.625
Iteration 5	0.656	0.18	0.577

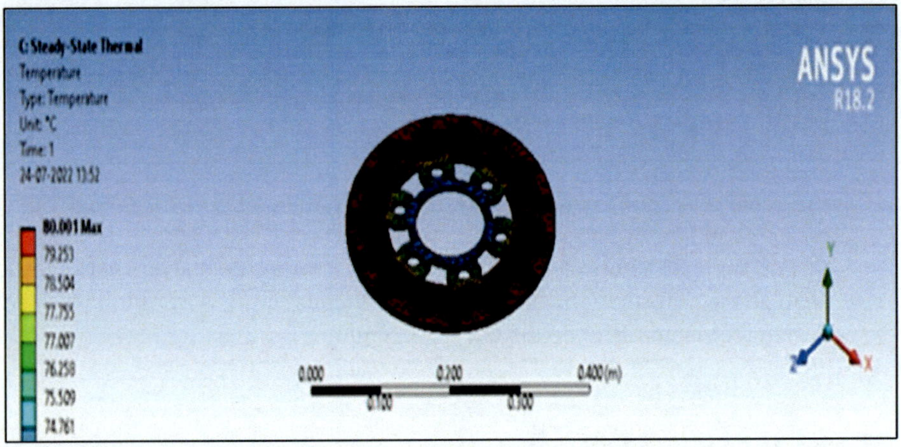

Fig. 5 Temperature distribution (iteration 4)

In Fig. 6, for Iteration 5, maximum and minimum total heat flux was observed to be 37,404 W/m^2 and 8314.1 W/m^2, respectively, while the lowest temperature observed was 76.635 °C.

Table 3 summarizes the results obtained for all the iterations of the disc brake models for steady-state thermal analysis (before topology optimization).

The following results were obtained for static structural analysis after topological optimization.

In Fig. 7, for Iteration 1, maximum von Mises stress and total deformation of 0.616 MPa and 0.193 microns were observed, respectively, during the static structural analysis.

Similarly, maximum von Mises stress and total deformation of 0.567 MPa and 0.182 microns for Iteration 2 and 0.574 MPa and 0.219 microns for Iteration 3 were observed, respectively, during the static structural analysis.

Table 4 summarizes the results obtained for all the iterations of the disc brake models for static structural analysis (after topology optimization).

The following results were obtained for static-thermal analysis after topological optimization.

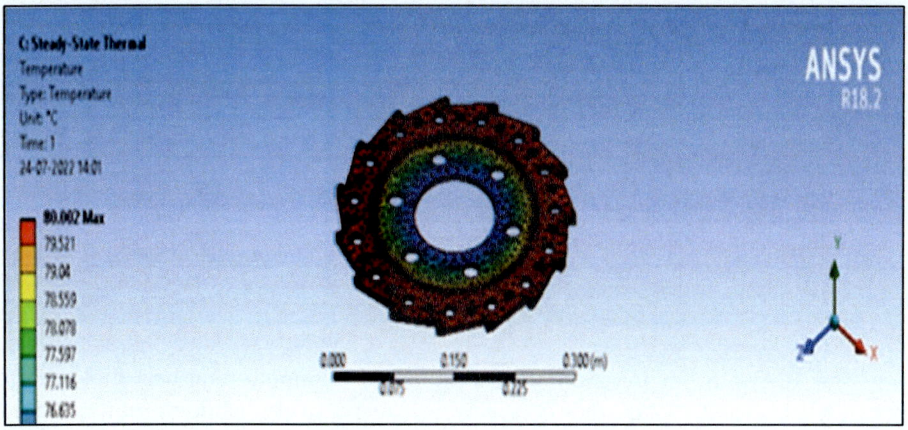

Fig. 6 Temperature distribution (iteration 5)

Table 3 Summary of steady-state thermal analysis (before topology optimization)

Parameter	Before topology optimization			
	Total heat flux (W/m^2)		Temperature (°C)	
	Maximum	Minimum	Maximum	Minimum
Iteration 1	38,555	8567.7	80	76.635
Iteration 2	43,117	9581.7	80	76.634
Iteration 3	38,123	8473.6	80	76.606
Iteration 4	46,137	10,253	80	74.761
Iteration 5	37,404	8314.1	80	76.635

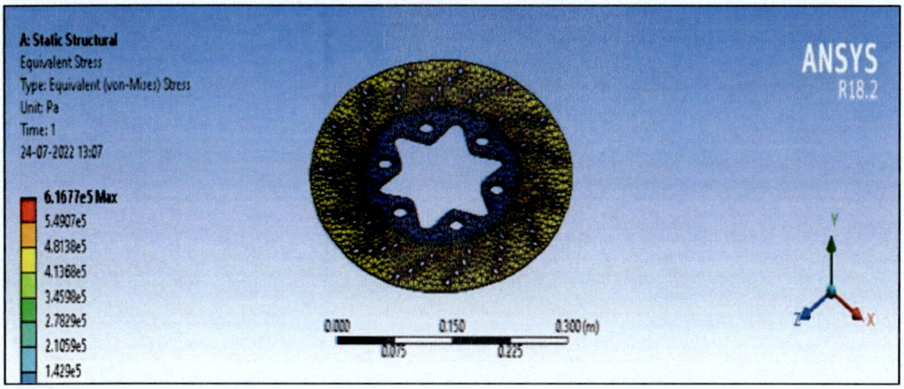

Fig. 7 von Mises stress (iteration 1)

Table 4 Summary of static structural analysis (after topology optimization)

Parameter	After topology optimization		
	Mass (kg)	Total deformation (microns)	von Mises stress (MPa)
Iteration 1	0.683	0.193	0.616
Iteration 2	0.661	0.182	0.567
Iteration 3	0.488	0.219	0.574
Iteration 4	0.665	0.2	0.605
Iteration 5	0.592	0.182	0.574

In Fig. 8, for Iteration 4, maximum and minimum total heat flux was observed to be 31,999 W/m^2 and 7110.9 W/m^2, respectively, while the lowest temperature observed was 76.554 °C. Similarly, for Iteration 5, maximum and minimum total heat flux was observed to be 30,291 W/m^2 and 6732.9 W/m^2, respectively, while the lowest temperature observed was 77.487 °C.

Table 5 summarizes the results obtained for all the iterations of the disc brake models for steady-state thermal analysis (after topology optimization).

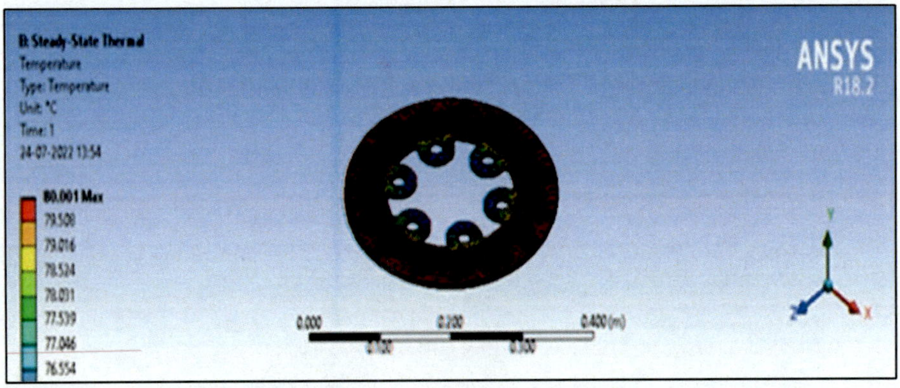

Fig. 8 Temperature distribution (iteration 4)

Table 5 Summary of steady-state thermal analysis (after topology optimization)

Parameter	After topology optimization			
	Total heat flux (W/m^2)		Temperature (°C)	
	Maximum	Minimum	Maximum	Minimum
Iteration 1	27,895	6219	80	77.487
Iteration 2	28,346	6299.2	80	77.486
Iteration 3	45,643	10,146	80	76.961
Iteration 4	31,999	7110.9	80	76.554
Iteration 5	30,291	6732.9	80	77.487

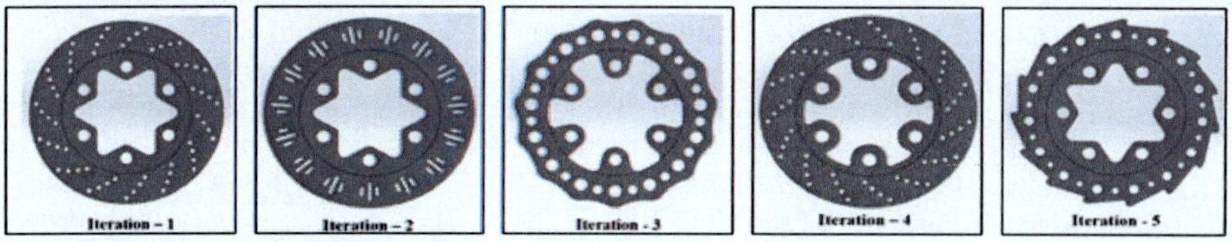

Fig. 9 3D models of different topology-optimized disc brake iterations

The data showcased in Tables 2 and 4 indicate the various values of stress and deformation obtained for before and after topology optimization for different disc brake iterations. Percentage (%) reductions of 8.57%, 8.83%, 1.21%, 2.78%, and 9.76%, respectively, were obtained for different iterations. As it can be inferred from the above table, the % reduction in mass is significant for iterations 1, 2 and 5. But amongst them, Iteration 5 is better, owing to lesser von Mises stress induced and highest % reduction in mass. The data showcased in Tables 3 and 5 indicate that there is negligible change in the magnitudes of temperature for before and after topology optimization cases concluding that the topology optimization did not degrade the thermal performance of the disc brakes while optimizing the mass distribution of the disc brakes.

4 Conclusion

In this paper, topology optimization of various disc brake designs was done, as indicated in Fig. 9. Static structural and steady-state thermal analysis was done in order to validate the optimized designs. A reduction of 10% in mass was observed in one of the disc brake iterations. Also, negligible deviations were observed in the structural and thermal performance of the disc brakes. For future scope, researches can study the effect of different materials on the structural and thermal performance of the disc brakes subjected to topology optimization.

References

1. Umale SR, Varma D (2016) Analysis and optimization of disc brake rotor. IRJET 3(11). e-ISSN: 2395-0056
2. Matariya V, Patel H (2019) Topological optimization of automobile rotor disk brake. SSRG Int J Mech Eng 6(4):23–27
3. Eswaraiah V, Govindu J (2021) Design optimization of disk brake by using finite element analysis. Int J Inno Res Technol (IJIRT) 8(4), Sept 2021
4. Ranjith Kumar G, Thriveni S, Rajasekhar Reddy M, Harinath Gowd G (2014) Design analysis & optimization of an automotive disc brake. Int J Adv Eng Res Sci (IJAERS) **1**(3)
5. Thigale S, Shah C. Weight reduction in brake disc using topology optimization. IJRET: Int J Res Eng Technol. e-ISSN: 2319-1163 | p-ISSN: 2321-7308

6. Naikwadi A, Wadageri CS, Bidari KY (2017) Design optimization of disc brake rotor. Int Res J Eng Technol (IRJET) 4(9). e-ISSN: 2395-0056
7. Pachpore SS, Jadhav PV, Ghorpade RR (2021) Analyzing relation of canal operating force and canal curvature in RCT: A mathematical overview. Mater Today: Proc 47:5690–5696. https://doi.org/10.1016/j.matpr.2021.03.746
8. Ghatti M, Ghatole K, Pachpore SS (2021) Optimization in spring design using knowledge-based engineering approach. AIP Conf Proc 2358(1). AIP Publishing LLC
9. Thakur AS, Dhakad PS (2018) Thermal analysis of disc brake using ANSYS. Int J Tech Inno Mod Eng Sci (IJTIMES) 4(6), e-ISSN: 2455-2585
10. Bhandari VB. Machine design data book. Mc Graw Hill Publication
11. Gajinkar AD et al (2021) Topology optimization of foot pedal using generative design approach. New Arch-Int J Contemp Arch 8(2), 1476–1491
12. Galande A, Pachpore SS, Jadhav PV, Ghorpade R, Lele MM (2023) Investigation and assessment for specific volume of Gutta-Percha as a biomaterial in RCT. Mater Today: Proc 72:741–747. https://doi.org/10.1016/j.matpr.2022.08.499
13. Pachpore SS, Botre MK, Patil AS, Jadhav PV (2020) Development and validation of transportation methodology by predicting dynamic behavior of container for safe transportation. InTechno-Societal 2018: Proceedings of the 2nd International Conference on Advanced Technologies for Societal Applications-Volume 1, pp 943–953. Springer International Publishing

Applications and Limitations of Additive Manufacturing Techniques for Manufacturing Components of Aerospace Industry

Krish V. Raibole and Sachin R. Deshmukh

Abstract The present paper looks into the types of additive manufacturing relevant to aerospace industry while discussing its advantages and limitations with future fields requiring research and development, while also discussing the impact of additive manufacturing (AM) on the manufacturing industry and the reasons due to which it has remained growing swiftly in numerous fields in the past couple of years. Amongst them, aerospace has been attracting further attention due to substantial funding by the primary aeronautical firms for evolving the AM industrial applications. Nevertheless, a lot of research and development has been progressing to make it more adaptable and a harmless technology and needs making development in original materials, technologies, process design and cost efficiency.

Keywords Additive manufacturing · Aerospace · Industry · Adaptable · Cost efficiency

Nomenclature

AM	Additive manufacturing
SLM	Selective laser melting
DMLS	Direct metal laser sintering
SLM	Selective laser melting
EBM	Electron beam melting
GE	General electric
LEAP	Leading-edge aviation propulsion
NDE	Non-destructive evaluation
BMG	Bulk metallic glasses
NASA	National Aeronautics and Space Administration

K. V. Raibole · S. R. Deshmukh (✉)
Department of Mechanical Engineering, Dr. Vishwanath Karad MIT World Peace University, Pune 411038, India
e-mail: sachin.deshmukh@mitwpu.edu.in

N. Kumar et al. (eds.), *Advances in Materials and Agile Manufacturing*,
Lecture Notes in Mechanical Engineering,
https://doi.org/10.1007/978-981-99-6601-1_2

1 Introduction

Additive manufacturing (AM) is a method involving material joining to achieve 3D parts with required geometries, generally done by adding one layer after another as opposed to formative manufacturing methodologies and subtractive manufacturing processes [1]. The primary advantage of AM in production applications for the aerospace industry is the capability to generate intricate engineered geometries with a restricted number of processing steps with a comparative low production cost [2]. In recent years, with development of new techniques in AM industry, aerospace manufacturing industry highly prefers AM in fields of design, research and prototyping as it is suitable for low production volumes while reducing lead time and cost of production in rapid prototyping. According to the Airbus Group, one of the leading companies in the aerospace industry, the direct effect and advantage of adopting AM technology is a cost reduction between 30 and 55% [3]. Methods such as selective laser melting (SLM) allow a layer-by-layer production of lightweight structures of titanium alloys for aircraft fuselage and wings which result in an increase of efficiency along with reduced fuel consumption [4].

2 Classification of AM

2.1 Power Bed Fusion Process

(a) **Direct Metal Laser Sintering (DMLS)**

DMLS technology makes use of a high-powered laser or multiple lasers to sinter, melt and solidify individual layers of metal powder. This technology uses the primary method of sintering which is a thermal procedure of altering loose fine particles into a solid coherent mass by heat and/or pressure without completely melting the particles to the melting point. This technique is highly favourable for the aerospace industry as it is known to produce completely dense components with extreme precision in a comparatively less time [5].

(b) **Selective Laser Melting (SLM)**

This process involves a sealed compartment occupied with an inert gas like Argon, wherein a fabrication piston lowers the construction plate while another piston drives up the powder so that a roller can finely dispense a coating on top of the object. After which a laser is produced and focused through a scanner towards the powder layer which shapes the fused layer as per the requirement of the designed component, a wide variety of metallic materials can be optimized in the SLM process, although these materials need to pass certain criteria such as grain size and their chemical composition. Figure 1 shows the classification of SLM under Power bed fusion processes along with other types of metal based additive manufacturing processes [6]. Figure 2 displays the order of operations executed in conventional SLM [7].

Range in which SLM can be applied:

1. Grain size: < 100 μm
2. Median grain size: 45–63 μm
3. Layer thickness: 10–100 μm.

Although the range for layer thickness is between 10 and 100μm, in order to achieve higher part resolution and surface quality, the recommended layer thickness should be of 30 μm [7].

(c) **Electron Beam Melting (EBM)**

The major difference in SLM and EBM is the heat source; EBM makes use of an electron beam while SLM uses a laser beam. In order to circumvent residual stresses in the finished product, which might negatively affect the mechanical properties, EBM adjusts the process temperature so that it remains continually high in the chamber, particularly above the stress-relieving temperature, for the entirety of the process [7]. In EBM, components are fabricated in a vacuum as the electron beam requires a vacuum to operate in. Maximum circular build diameter and height that can be manufactured using EBM are 350 mm and 380 mm, respectively [5].

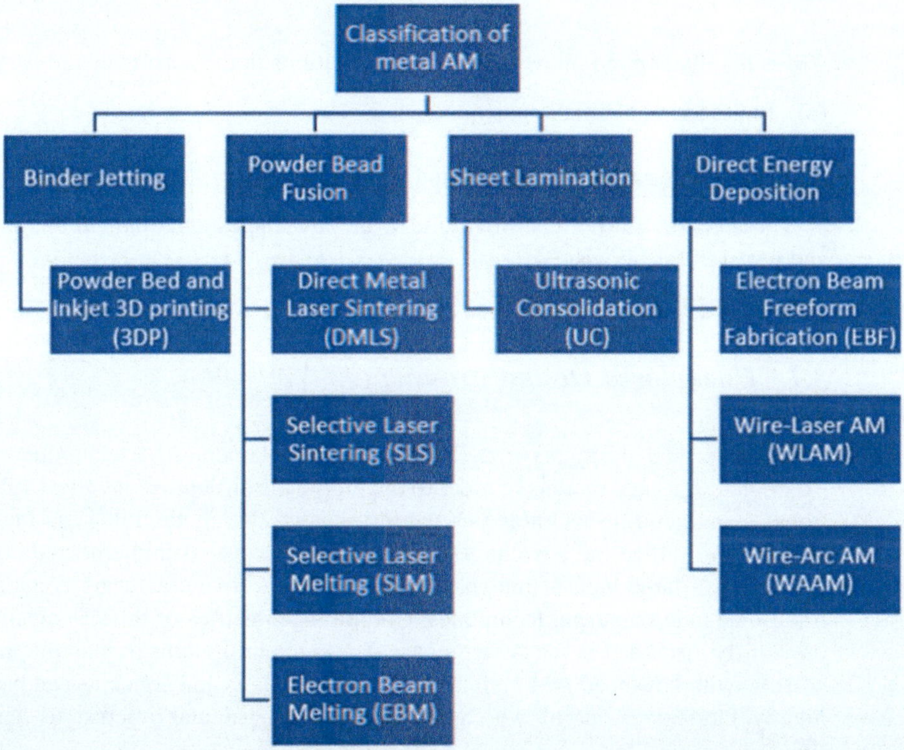

Fig. 1 Classification of metal-based AM [6]

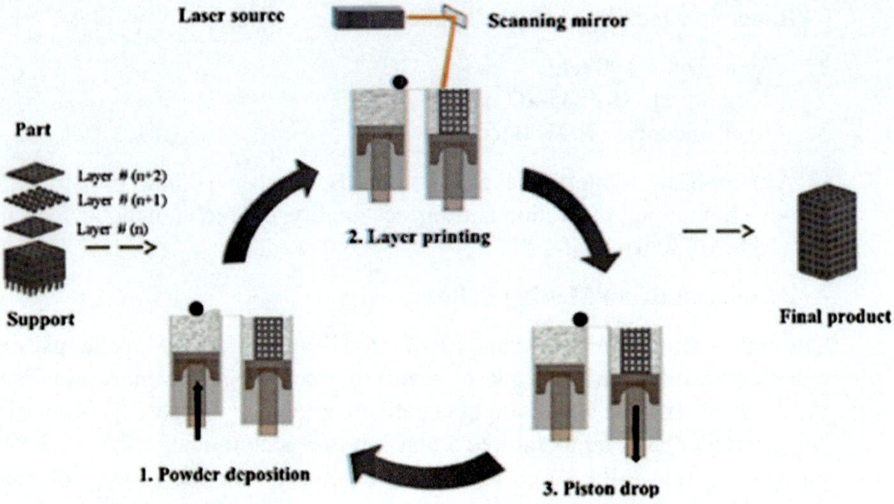

Fig. 2 A schematic of operations performed in SLM [7]

3 Applications of AM in Aerospace Industry

AM is mainly applied in the following applications in the aerospace industry:

(a) Engineering of small-scale models
(b) Production of prototypes
(c) Manufacturing of spare parts and components [8].

These can be further classified under the advantages and applications in design and research as follows.

3.1 Unexplored Design Frontiers

Researchers and designers generally tend to implement AM techniques to pre-existing designs and models in order to overcome the difficulties involved with traditional manufacturing technologies. Unfortunately, this methodology is ineffective as it fails to utilize the advantages AM offers. However if implemented correctly, AM opens a large area of unexplored design due to manufacturing constraints of traditional manufacturing techniques. Complex geometries or intricate design were previously discarded in the design phase due to the limitations in manufacturing.

Now with advanced AM methods, such new designs and concepts can be conveniently fabricated, out of which several are of particular interest to aerospace [9].

I. Gradients

Gradient materials are a type of material that bear different properties at different geometric locations along its characteristic length and serve as a solution to many engineering problems [9]. In space shuttles or rockets, materials which make up the engine nozzles require high-temperature tolerance varying along its length; hence, an alloy going from titanium to niobium has been conceived for such an application [10].

II. Ceramics

Ceramics is a type of material which is comparatively easier to manufacture using AM, and it has one of the most suitable properties; especially for space flight, it can be used in extreme temperature and pressure applications, piezoelectric and electrochemical devices. Components such as ceramic biomaterial flexures toughened surfaces with a ductile interior can be fabricated. Ceramics having dielectric and magnetic properties varying along its characteristic length can also be fabricated as graded components [9].

III. Microtrusses

Although AM can only fabricate microtrusses as small as few micrometres, whereas using conventional methods microtrusses can be made on a molecular scale, it is very convenient to fabricate microtrusses having extreme strength to mass ratios using AM [9–11].

3.2 Combination of Multiple Assemblies

Intricate components such as brackets, trusses, nozzles and couplings having numerous components requiring assemble post manufacturing cannot be manufactured collectively as a sole component using conventional machining techniques. However, with the induction of AM technology, such parts can be manufactured as a single component resulting in reduction of the total amount of parts. Previously unavoidable elements such as fasteners, nuts, bolts, rivets, needed for assembly are deemed unnecessary by AM [9].

GE Aviation implements the direct metal laser sintering process to produce their succeeding generation jet engine components. Each LEAP engine within it has 19 fuel nozzles created via 3DP, an advanced AM technique developed by ExOne. The design elements include supplementary intricate cooling pathways and support ligaments, which result in five times the service life when equated with the precursor made by orthodox manufacturing. AM process permitted engineers to use an easier design that decreases the quantity of brazes and welds from 25 to only 5; the quantity of parts used to make the nozzle was lessened from 18 to 1. In all, the weight of these nozzles is 25% lighter than that of the predecessor [12–15].

3.3 Mass Reduction

Weight and/or mass reduction in material components is mainly attained through topology optimization in AM. This method provides a mathematical design resolution so that there is efficient dispersal of material for a given set of boundary circumstances, loading conditions and physical space design. AM implements topology optimization in such a way so that it can produce lightweight structural parts which cannot be conventionally manufactured [16].

Weight reduction in aerospace industry is of utmost importance at it leads to reduction in the total mission cost, and if the weight of the spacecraft is as minimum as conceivable, the mass ratio upsurges, and thus, the Δv budget rises for a certain volume of fuel. Lower mass even at the time of launch of rockets and space shuttles likewise cuts down the expense of the entire mission significantly [17]. As mentioned earlier, GE Aviation manufactured an engine with 19 fuel nozzles with the use of AM, in which the nozzle fabricated using AM technologies was 25% lighter when compared to its precursor [12]. Similarly, a titanium bracket connector used in the Airbus A350 XWB manufactured using a laser-based AM method was 30% lighter than its predecessor made from casting aluminium [12].

4 Limitations of AM in Aerospace Industry

4.1 Standardization

Although AM provides attractive advantages with respect to production time and cost against conventional methods, components fabricated using AM must be able to fulfil the requirements, certifications and standardizations laid out by international safety agencies such as the European Aviation Safety Agency and the Federal Aviation Administration. Only when the components achieve all the required certifications, they can be equipped onboard any aircraft or spacecraft. Even AM parts which are used for repair or maintenance must be developed by an approved design organization or an organization with comparable abilities [18]. In this aspect, traditional manufacturing for metallic components has an advantage over AM as these methods have already well-established standards and regulations, implying that protocols for manufacturers and suppliers are also recognized globally.

Efforts are being made at present by international boards and teams globally to develop standardization for AM methods which will enable the recognition of AM terminology, material, process, maintenance and safety procedures [13–18].

4.2 *Non-destructive Evaluation (NDE)*

At the moment, there is a severe lack of non-destructive evaluation (NDE) processes integrated into AM, which poses as a major hindrance for the reception of AM fabricated parts in NASA and other aerospace establishments. NDE methods are currently being applied to traditional metal manufacturing, and hence, there is a deficiency of data available for NDE of AM. This points that there is a need to compile supplementary NDE data notably by starting material, process method and NDE method. This approach will surge NDE experience and information applicable to AM. Non-destructive evaluation can conceivably be used to portray microstructure and morphology of input materials through processing and post-processing. In addition to defect tracking, finer-scale atomic, molecular and chemical configurations can also be measured by NDE throughout the part's lifespan. Assurance in the ability to guarantee AM part quality by non-destructive evaluation is also a matter of concern due to the prospect of generating defects during processing that are minor than which can be distinguished [7, 8].

Standardization of NDE for AM will be only possible if sufficient data on AM is available, after which the general goal will be to comprehend the types of naturally occurring defects produced by the AM process, what their effects are and what NDE techniques are most appropriate for their discovery. Qualification and certification of NDE-based protocols for flight hardware must be developed which will rely on testing and modelling and which will lead to greater acceptance of AM into aerospace industry [18].

4.3 *Choice of Materials*

Choice and selection of materials play a significant part in defining the final weight and performance of an aircraft [6, 5]. AM manufactured parts generally tend to have deprived mechanical properties when equated to those fabricated by means of conventional methods [6]. Furthermore, materials manufactured via AM develop enduring stresses which are focused towards the root of the additively manufactured part, near the boundary with the substrate. Additionally, the longitudinal stress components (parallel to the welding direction) are perceived to be foremost, with stress levels surpassing 50% of the materials' yield strength [1–8, 19, 20]. Nevertheless, if buckling throughout build is circumvented, stresses can be dealt with by a comparatively rapid heat treatment [3]. Titanium, aluminium, nickel, stainless steel, tool steel, etc., are frequently used in AM for aerospace manufacturing. AM offers decrease of material wastage, total cost of arranging of waste material and material conveyance cost, resulting in lessening of net product costs [5, 6, 20].

While AM could produce products from an extensive variety of plastics and metals, the raw material choice is limited. This is because not all plastics or metals can be temperature controlled to permit for 3D printing or other AM methods. Primarily, amorphous or glassy alloys frequently called bulk metallic glasses (BMGs) are hopefully a revolution in the applications capable of applying their unparalleled pliability in AM fabrication [8].

5 Conclusion

This paper encompasses all the advantages and disadvantages AM methodology has to offer to the aerospace industry; despite being vastly advanced when compared to current traditional subtractive manufacturing techniques, its application in the aerospace industry is as of now very limited, primarily due to lack of standardization and safety certifications for AM processes and materials as discussed in detail above.

Scope for further studies includes work mainly to be done in the field of regulating AM methods for them to be approved by international safety boards globally, allowing AM to contend in the manufacturing marketplace. One key advantage provided by AM is that components fabricated using AM are greatly lighter than its precursors. With the increasing demand of the aeronautical and aerospace industry, AM could revolutionize the manufacturing industry by adapting to its requirements better than traditional methods. This could lead to the entire aerospace industry relying completely on AM for any and all kinds of fabrication, from small brackets and rivets to entire aircrafts being additively manufactured.

References

1. Altıparmak SC, Xiao B (2021) A market assessment of additive manufacturing potential for the aerospace industry. J Manuf Process
2. Gibson I, Rosen D, Stucker B (2015) Additive manufacturing technologies 3d printing, rapid prototyping, and direct digital manufacturing, 2nd Ed., pp 468–472
3. Morcillo-Bellido, Martínez-Fernández J, Morcillo-García (2019) Impact of additive manufacturing in aerospace industry purchasing process. In: 13th International conference on industrial engineering and industrial management
4. Uhlmanna E, Kerstinga R, Kleina TB, Cruzb MF, Borille AC (2015) Additive manufacturing of titanium alloy for aircraft components. In: 15th Machining innovations conference for aerospace industry
5. Blakey-Milner B, Gradl P, Snedden G, Brooks M, Pitot J, Lopez E, Leary M, Berto F, du Plessis A (2021) Metal additive manufacturing in aerospace: a review. Mater Des
6. Li JLZ, Alkahari MR, Rosli NAB, Hasan R, Sudin MN, Ramli FR (2019) Review of wire arc additive manufacturing for 3d metal printing. Int J Autom Technol
7. Gisarioa A, Kazarianb M, Martinac F, Mehrpouya M (2019) Metal additive manufacturing in the commercial aviation industry: a review. J Manuf Syst
8. Oyesolaa M, Mpofa K, Matheb N (2019) A techno-economic analytical approach of laser-based additive manufacturing processes for aerospace application, In: 2nd International conference on sustainable materials processing and manufacturing (SMPM 2019)
9. Shapiro AA, Borgonia JP, Chen QN, Dillon RP, McEnerney B, Polit-Casillas R, Soloway L (2016) Additive manufacturing for aerospace flight applications. J Spacecraft Rock

10. Hofmann DC, Kolodziejska J, Roberts S, Otis R, Dillon RP, Suh JO, Liu ZK, Borgonia JP (2014) Compositionally graded metals: a new frontier of additive manufacturing. J Mater Res 29(17)
11. Zheng X, Lee H, Weisgraber TH, Shusteff M, DeOtte J, Duoss EB, Kuntz JD, Biener MM, Ge Q, Jackson JA, Kucheyev SO, Fang NX, Spadaccini CM (2014) Ultralight, ultrastiff mechanical metamaterials. Science 344(6190)
12. Liu R, Wang Z, Sparks T, Liou F, Newkirk J (2017) Aerospace applications of laser additive manufacturing. In: Laser additive manufacturing
13. Kellner T (2014) Fit to print: new plant will assemble world's first passenger jet engine with 3d printed fuel nozzles. In: Next-gen materials
14. Kamal M, Rizza G (2019) Design for metal additive manufacturing for aerospace applications. In: Additive manufacturing for the aerospace industry
15. Jyothish Kumar L, Krishnadas Nair CG (2017) Current trends of additive manufacturing in the aerospace industry. In: Advances in 3D printing & additive manufacturing technologies
16. Orme ME, Vernon R, Madera IJ, Yancey R, Mouriaux F (2017) Additive manufacturing of lightweight, optimized, metallic components suitable for space flight. J Spacecraft Rockets
17. Sacco E, Ki Moon S (2019) Additive manufacturing for space: status and promises. Int J Adv Manuf Technol
18. Uriondo A, Esperon-Miguez M, Perinpanayagam S (2015) The present and future of additive manufacturing in the aerospace sector: a review of important aspects. J Aerosp Eng
19. Tepylo N, Huang X, Patnaik PC (2019) Laser-based additive manufacturing technologies for aerospace applications. J Adv Eng Mater
20. Barroqueiro B, Andrade-Campos A, Valente RAF, Neto V (2019) Metal additive manufacturing cycle in aerospace industry: a comprehensive review. J Manuf Mater Process

Optimization of WEDM Process Parameters Using Grey-Taguchi Approach for the Development of Turning Tool Insert

Vishwajeet Kumar and Subhas Chandra Mondal

Abstract This research involves the development of the turning tool insert using an optimal combination of wire electrical discharge machining (WEDM) process parameters. It is very challenging to get high productivity and good surface finish simultaneously in manufacturing industries. However, grey relational analysis (GRA) is one of the simple and easy-to-implement multi-objective optimization techniques. The input parameters have been selected as pulse on time, pulse off time, current and voltage for machining of Fe-based hardfaced deposit. Material removal rate (MRR) and surface roughness (SR) have been selected as output parameters. The optimal combination of machining process parameters has been evaluated using GRA optimization technique for high productivity and good surface finish. Current has a predominant role during the machining while voltage has the least impact on response characteristics.

Keywords Machining · WEDM · GRA · MRR · SR · Pulse on time · Pulse off time · Current · Voltage

1 Introduction

Wire electrical discharge machining (WEDM) is widely used to machine composite, ceramics, super alloys, carbides, tool steel, stainless steel and heat-resistant steel by generation of a series of sparks between tool and workpiece in WEDM process which leads to a wide range of applications in various manufacturing industries for the machining of [1, 2]. It is very challenging for a WEDM process to machine at a high production rate with good surface integrity. However, these types of problems have been resolved by several researchers using different multi-criteria decision-making techniques such as technique for order of preference by similarity to ideal

V. Kumar · S. C. Mondal (✉)
Department of Mechanical Engineering, Indian Institute of Engineering Science and Technology, Shibpur, Howrah, West Bengal 711103, India
e-mail: scmondall@gmail.com

N. Kumar et al. (eds.), *Advances in Materials and Agile Manufacturing*, Lecture Notes in Mechanical Engineering,
https://doi.org/10.1007/978-981-99-6601-1_3

solution (TOPSIS), multi-objective MOPSO, entropy method, complex proportional assessment (COPRAS), evaluation based on distance from average solution (EDAS), additive ratio assessment (ARAS) and multi-objective optimization by ratio analysis (MOORA).

Surface roughness and MRR have been studied by different modelling and optimization techniques for various materials. Kung and Chiang [3] reported the machinability of ceramics (Al_2O_3 and TiC). Pramanick et al. [4] studied the effect of the process parameters of WEDM for the machining of spark plasma sintered boron carbide specimens using the ANFIS technique. Patel et al. [5] introduced artificial neural networks with a backpropagation and response surface methodology including central composite design for the modelling and prediction of MRR in the WEDM process. Wasif et al. [6] studied the machinability of Al 5454 alloy by applying genetic algorithm to increase productivity. Hammami et al. [7] reported the pulse on time and off time as the significant cutting parameters during the machining of aluminium alloy for surface integrity. Naveed et al. [8] simultaneously optimized geometrical errors at the top as well as bottom surfaces and Ra using the grey relational analysis (GRA). The surface roughness has been enhanced by 12%. Saha and Mondal [9] reported the machining of nano-structured hardfaced using principal component analysis. Hardfaced components are very difficult to machine using a conventional machining process due to their high hardness, adequate fracture strength and exceptional wear resistance. However, the machining of hardfaced deposits can be performed efficiently using the non-traditional machining process.

Several researchers and practitioners have reported the machining of composite, ceramics and different grades of steel using WEDM, although very few research papers are available for the machining of Fe-based hardfaced deposit. Therefore, an attempt has been made to machine Fe-based hardfaced deposit using WEDM for the development of the turning tool insert. Material removal rate and surface roughness have been selected as output parameters while pulse on time, as pulse off time, current and voltage have been selected as input parameters. The optimization of wire electrical discharge machining (WEDM) process parameters has been performed using GRA. The optimal combination of machining process parameters has been used for the development of the turning tool insert.

2 Experimentation

2.1 *Materials and Equipment*

Electronica Eurocut Mark I CNC WEDM was used to conduct the experiments. Fe-based hardfaced consumables were deposited on the surface of mild steel using manual metal arc welding. Hardfaced deposit is widely used for blanking and stamping dies, punches and rotary shear blades. Brass wire of 0.25 mm diameter was used as the electrode while deionized water was used as a dielectric medium

Table 1 Composition of base metal and hardfaced deposit

Content	C	Mn	Si	Cr	W	V	Mo	P	S	Fe
Base material	0.2–0.23	1.50	0.45	–	–	–	–	0.04	0.05	Rest
Hardfaced deposit	0.8–1	–	–	4–5	2–3	1–2	8–9	–	–	Rest

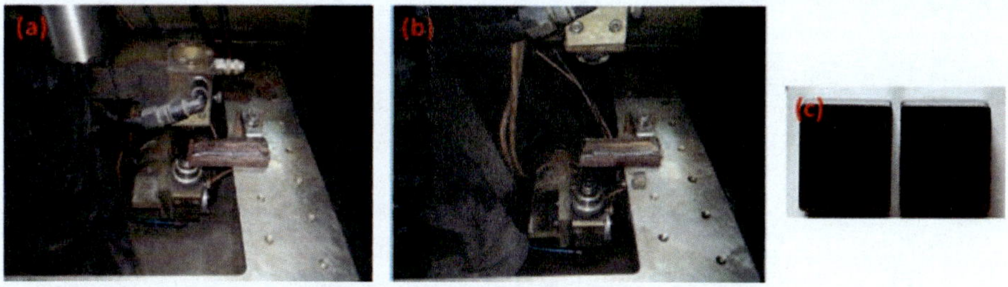

Fig. 1 **a** Machining of Fe-based deposit, **b** WEDM machined specimen and **c** Fe-based hardfaced inserts

to perform the machining of Fe-based hardfaced deposit at the different sets of machining process parameters. Each experiment was carried out with a constant length of cut (12 mm). Table 1 shows the composition of base metal and hardfaced deposit. Figure 1 shows the machining of Fe-based hardfaced deposit.

2.2 Measurement of Response Characteristic

The machining efficiency of WEDM is usually evaluated in terms of MRR. MRR was evaluated with the help of the machining speed and thickness of hardfaced workpiece using the following equation:

$$\text{MRR}(\text{mm}^2/\text{min}) = M_S(\text{mm}/\text{min}) * T_C(\text{mm}) \tag{1}$$

where M_S = machining speed shown on the panel, and T_C = thickness of machining specimen. WEDM generates microscopic cracks, craters, uneven hills and valleys on the machined surfaces. The centre line average (Ra) value of machined surfaces was measured three times for each experiment. The average value of experimental data for Ra was considered in this study.

Table 2 Design matrix for machining process parameters

S. No.	Impulse current, Ip (A)	Pulse on time, T_{on} (μs)	Pulse off time, T_{off} (μs)	Servo voltage, SV (V)
1	80	10	40	30
2	80	15	45	40
3	80	20	50	50
4	80	25	55	60
5	100	10	45	50
6	100	15	40	60
7	100	20	55	30
8	100	25	50	40
9	120	10	50	60
10	120	15	55	50
11	120	20	40	40
12	120	25	45	30
13	140	10	55	40
14	140	15	50	30
15	140	20	45	60
16	140	25	40	50

3 Design of Experiment

The design of experiment is a scientific approach, which is used to carry out the modelling and analysis of the influence of process variables on the response variables [10]. Taguchi L16 orthogonal array was used to design the machining parameters of WEDM. Four machining parameters like impulse current, pulse on time, pulse off time and servo voltage were selected for the machining of Fe-based hardfaced deposit. Four factors and four levels of machining parameters were chosen in this study. The design matrix for machining process parameters and their levels have been shown in Table 2.

4 Grey Relational Analysis (GRA)

Dr. Deng has introduced the concept of grey theory using information, which is unsure and incomplete. It uses the available information for the formulation of a mathematical model to predict incomplete information. The mathematical model recognizes each response characteristic. It establishes a correlation between experiential data. The following steps are involved in the GRA optimization technique:

1. Normalize the experimental results of MRR and SR Using Eqs. (2) and (3), respectively.

(a) For larger the better criterion

$$N_{ij} = \frac{E_{ij} - \min(E_{ij})}{\max(E_{ij}) - \min(E_{ij})} \tag{2}$$

(b) For lower the better criterion

$$N_{ij} = \frac{\max(E_{ij}) - E_{ij}}{\max(E_{ij}) - \min(E_{ij})} \tag{3}$$

where N_{ij} and E_{ij} are normalized data and experimental data, respectively.

2. Calculate the absolute difference values.

$$\Delta_{ij} = \text{mod}(N_0 - N_{ij}) \tag{4}$$

3. Calculate the Grey Relational Coefficient (GRC).

$$\text{GRC}_{ij} = \frac{\Delta_{\min} + \Psi \times \Delta_{\max}}{\Delta_{ij} + \Psi \times \Delta_{\max}} \tag{5}$$

where Ψ represents the distinguishing coefficient. To provide equal weight to each response characteristic, the distinguishing coefficient was set to 0.5.

4. Calculate the grey relational grade (GRG) by averaging the GRC values of individual experimental data.
5. Choose the best values of process parameters and validate them with a confirmation test.

5 Results and Discussions

Machining of Fe-based hardfaced deposit has been performed with the help of WEDM. The experimental data has been evaluated using L16 orthogonal array. The performance characteristics (MRR and surface roughness) have been optimized with the help of GRA technique. Table 3 represents experimental data for MRR and surface roughness, normalized data, deviation sequence, GRC and GRG. Data preprocessing is the first step for GRA. It removes the impact of unit and predominance of one data sequence over the others by converting the experimentally observed data in the range 0 to 1 [11]. Depending on the characteristics of the data sequence, the higher the better criteria have been applied for the MRR, while the lower the better criteria have been chosen for surface roughness. The normalized value is further utilized for the determination of variation of data from the ideal solution using Eq. (4). Grey relational coefficient (GRC) represents the relational degree between the ideal solution and normalized data. It has been calculated using Eq. (5). Finally, the average value of GRC of corresponding MRR and Ra provides the grade of each experiment. It is represented by GRG, which gives information regarding the interaction of response characteristics.

Figure 2 depicts the variation of GRG Data with WEDM process parameters. It represents the correlation between performance characteristics. The higher value of the grey relational grade indicates a good correlation between performance characteristics [12]. It has been observed that the combination of 80 A impulse current, 10 μs T_{on}, 40 μs T_{off} and 50 V servo voltage has provided the high value of GRG. Table 4 shows the mean value of GRG data for different combinations of WEDM process parameters. It shows that the impulse current has predominant role during the machining of Fe-based hardfaced deposit followed by T_{off}, and T_{on}. However, the SV has the least impact on the GRG.

Table 5 represents the response table for the confirmation test. It has been observed that the error is – 4.44% for GRG. However, the deviation of the actual experimental result from the predicted value has been observed 4.81% and 5.12% for MRR and Ra, respectively. The combination of 80 A impulse current, 10 μs T_{on}, 40 μs T_{off} and 50 V servo voltage has been used for the development of SNMN 120408 (ISO specification) turning tool insert. It has been shown in Fig. 1c. MRR and Ra values

Table 3 Experimental, normalized, deviation sequence, GRC and GRG data

S. No.	MRR (mm^2/min)	Ra (μm)	N_{MRR}	N_{Ra}	Deviation sequence for MRR	Deviation sequence for Ra	GRC for MRR	GRC for Ra	GRG
1	15.483	1.942	0.000	1.000	1.000	0.000	0.333	1.000	0.667
2	19.368	2.145	0.177	0.908	0.823	0.092	0.378	0.845	0.611
3	18.571	2.053	0.140	0.950	0.860	0.050	0.368	0.909	0.638
4	17.246	2.216	0.080	0.876	0.920	0.124	0.352	0.802	0.577
5	26.233	2.598	0.489	0.704	0.511	0.296	0.495	0.628	0.561
6	25.788	2.863	0.469	0.584	0.531	0.416	0.485	0.546	0.515
7	23.145	2.679	0.348	0.667	0.652	0.333	0.434	0.600	0.517
8	28.323	3.572	0.584	0.264	0.416	0.736	0.546	0.405	0.475
9	24.768	2.487	0.422	0.754	0.578	0.246	0.464	0.670	0.567
10	26.825	2.763	0.516	0.629	0.484	0.371	0.508	0.574	0.541
11	34.921	3.901	0.884	0.116	0.116	0.884	0.812	0.361	0.586
12	31.862	3.654	0.745	0.227	0.255	0.773	0.662	0.393	0.528
13	29.351	2.742	0.631	0.639	0.369	0.361	0.575	0.581	0.578
14	28.237	3.871	0.580	0.129	0.420	0.871	0.544	0.365	0.454
15	32.593	3.625	0.778	0.240	0.222	0.760	0.693	0.397	0.545
16	37.469	4.157	1.000	0.000	0.000	1.000	1.000	0.333	0.667

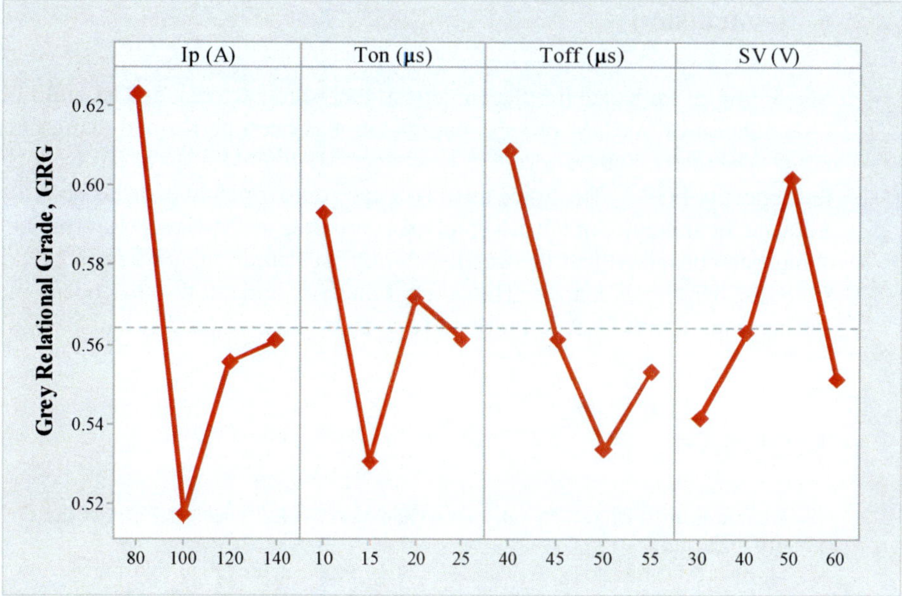

Fig. 2 Main effect plot for GRG

Table 4 Response table for mean of GRG

Level	Ip (A)	T_{on} (µs)	T_{off} (µs)	SV (V)
1	**0.6022**	**0.5808**	**0.6059**	0.5370
2	0.5062	0.5151	0.5466	0.5492
3	0.5466	0.5637	0.5241	**0.5910**
4	0.5568	0.5522	0.5351	0.5346
Delta	0.0960	0.0657	0.0818	0.0564
Rank	1	3	2	4

The bold values denotes the optimal combination of WEDM process parameters based on the high value of GRG

have been obtained as 19.422 mm^2/min and 1.746 µm, respectively at the optimum combination of WEDM process parameters.

Table 5 Response table for confirmation test

Output	Predicted value	Experimental value	Percentage error (%)
MRR (mm^2/min)	18.531	19.422	4.81
Ra (µm)	1.661	1.746	5.12
GRG	0.721	0.689	– 4.44

6 Conclusion

Machining of Fe-based hardfaced deposit has been performed. The multi-objective optimization of WEDM process parameters has been performed using GRA technique. The mathematical correlation between MRR and Ra has been developed using the concept of GRA. The grade (GRG) of each experimental data has been utilized to evaluate the influence of WEDM process parameters on MRR and Ra simultaneously. It has been observed that the impulse current has predominant role in machining of Fe-based hardfaced deposit. The optimal machining condition has been obtained at the combination of 80 A impulse current, 10 μs T_{on}, 40 μs T_{off} and 50 V SV.

References

1. Raghuwanshi BS (2009) Workshop technology, 9th Ed. Dhanpat Rai and Co. (P) Ltd., pp 919–922
2. Singh S, Maheshwari S, Pandey PC (2004) Some investigation into the electric discharge machining of hardened tool steel using different electrode materials. J Mater Process Technol 149:272–277
3. Kung KY, Chiang KT (2008) Modeling and analysis of machinability evaluation in the wire electrical discharge machining (WEDM) process of aluminum oxide-based ceramic. Mater Manuf Processes 23(3):241–250
4. Pramanick A, Mandal S, Dey PP, Das PK (2021) Comparative analysis for the prediction of WEDM responses for machining spark plasma sintered boron carbide ceramic sample by RSM and ANFIS. Mater Today: Proc 41:1089–1095
5. Patel S, Prajapati JM, Patel D, Patel M, Patel K (2022) Evaluation of different approach for WEDM process optimization. Mater Today: Proc 66:1988–1993
6. Wasif M, Khan YA, Zulqarnain A, Iqbal SA (2022) Analysis and optimization of wire electro-discharge machining process parameters for the efficient cutting of aluminum 5454 alloy. Alex Eng J 61(8):6191–6203
7. Hammami D, Louati S, Masmoudi N, Bradai C (2023) Influence of WEDM process parameters on aluminum alloy's surface finish. Int J Adv Manuf Technol:1–17
8. Naveed R, Ishfaq K, Harris M, Ahmed N (2023) WEDM of tapered rectangular geometry in tungsten–carbide cobalt composite (WC–Co): geometrical errors and surface roughness analysis. J Braz Soc Mech Sci Eng 45(1):1–16
9. Saha A, Mondal SC (2017) Experimental investigation and modelling of WEDM process for machining nano-structured hardfacing material. J Braz Soc Mech Sci Eng 39(9):3439–3455
10. Sahoo P, Routara BC, Bandyopadhyay A (2009) Roughness modelling and optimisation in EDM using response surface method for different work piece materials. Int J Mach Mach Mater 5(2–3):321–346
11. Lin JL, Lin CL (2002) The use of the orthogonal array with grey relational analysis to optimize the electrical discharge machining process with multiple performance characteristics. Int J Mach Tools Manuf 42(2):237–244
12. Chiang KT, Chang FP (2006) Optimization of the WEDM process of particle-reinforced material with multiple performance characteristics using grey relational analysis. J Mater Process Technol 180(1–3):96–101

The Application of the Preference Selection Index Method to Find the Best Work-Tool Combination During Machining

Sumit Mahajan and Kalyan Chakraborty

Abstract The medium carbon steel (MCS), Inconel 718, and D2 cold work steel were machined with different carbide tool materials. Speed, feed, and depth of cut were the machining parameters. The machining chips were collected for SEM (scanning electron microscopy) examination of chip surfaces. Tensile tests were done for all samples. Accordingly, strain-hardening exponent "n" and strength coefficient "K" were determined using the true stress-true strain graphs on the log–log graph papers. The chip reduction coefficient (CRC, ζ) and von Mises stress (VMS) were determined for different experimental conditions. The three different work-tool combinations were the alternatives, and CRC and VMS were the criteria. The preference selection index (PSI) method was applied to find out the best alternative. The fracture toughness of Inconel 718 was found to be higher among the test samples. The strain-hardening exponent of Inconel 718 was lower. The Inconel 718 and brazed carbide tool combination was ranked best, followed by D2 cold work steel and medium carbon steel. The result was validated by chip macrographs and SEM observation of chip surfaces.

Keywords Medium carbon steel · Speed · Strain hardening

1 Introduction

The PSI method can be applied to solve multi-criteria decision-making problems in machining. The main advantage with the use of the PSI method is that there is no necessity to determine the criteria weights by a separate method (e.g. the entropy method), as the PSI method contains the necessary step for the determination of criteria weights. The SiO2/SIC ceramic particulate reinforced aluminium alloy (AA2024) composite materials were prepared by stir casting. The tensile strength, flexural strength, impact strength, wear resistance, density, etc. were measured and

S. Mahajan · K. Chakraborty (✉)
M. E. Department, N.I.T Silchar, Silchar, Assam 788010, India
e-mail: kalyan@mech.nits.ac.in

N. Kumar et al. (eds.), *Advances in Materials and Agile Manufacturing*,
Lecture Notes in Mechanical Engineering,
https://doi.org/10.1007/978-981-99-6601-1_4

considered as criteria. The PSI method was applied to find the rank of the composites. The PSI method revealed that the composite with an equal number of reinforcing elements had the highest rank [1]. Titanium and nickel powder reinforced AA6061 composites were prepared by using the stir casting (high vacuum) method. By varying the weight per cent of titanium and nickel powders, various composites were created. The composite characterisation was done by measuring the tensile strength, flexural strength, impact strength, wear, density, void content, and hardness. The highest rank was attained by a composite having an equal amount of both powders as per the PSI method [2]. Biodegradable composites were prepared by reinforcing pine cone fibre and graphite in a polycaprolactone matrix. The measured objectives were the tensile strength, flexural strength, impact energy, fracture toughness, ductility, diffusion, sorption, permeability, and biodegradation (days). According to the PSI method, the composite with fibre (0%), matrix (100%), and graphite (0%) achieved the highest rank [3]. The objective of the present study was to (i) select the best combination of workpiece and tool among three different combinations by using the preference selection method and (ii) validate the best selection by qualitative assessment.

2 Experimentation

2.1 Material, Tool and Machining Parameters

Chemical compositions of three different workpiece materials (1. MCS, 2. Inconel 718, and 3. D2 cold work steel) are shown in Table 1. The tool geometry of two different tools is shown in Table 2. The experimental machining parameters are shown in Table 3.

Table 1 Workpiece number and chemical compositions of workpieces

Number	Chemical compositions
1	C:0.26, Mn:0.96
2	C:0.08, Mn:0.35, P:0.015, S:0.015, Si:0.35, Cr:0.21, Ni:0.55, Mo:3.3, Cb:5.5, Ti:1.15, Al:0.08, Co:1, B:0.06, Cu:0.3, Ta:0.05, Fe: Balance
3	C:1.5, Mn:0.3, V:0.9

Table 2 Tool geometry of tools

Tool type	Rake angle (deg.)	Clearance angle (deg.)	Main cutting edge angle ϕ, (deg.)
Carbide tip tool (Brazed), C1	4	7	34
Carbide tip tool (Brazed), C2	12	10	40

Table 3 Machining parameters

Parameter/Material	1	2	3
Speed, m/min	51.052	58.85	62.91
($f = 0.138$ mm/rev., $d = 1$ mm)			

2.2 *Methods*

The tensile tests for all the specimens were done using the universal tensile testing machine. The true stress and true strain data were estimated from the engineering stress and engineering strain data. The log–log graph paper was used to plot the true stress and true strain data (between the yield point and the ultimate strength point). A straight line was obtained by plotting the data. The slope of the line was the strain-hardening index "n". The point of intersection of the vertical line at true strain 1 is noted on the graph. The point of intersection of the horizontal line through this point with the y-axis was the strength coefficient. The chips were obtained for three different experimental conditions. The formed chip thicknesses were measured. The chip reduction coefficient for each experimental condition was given by Eq. 1:

$$\text{CRC} = \frac{t2}{t1}, \tag{1}$$

where $t2$ is the chip thickness and $t1$ is the uncut chip thickness

The uncut chip thickness $t1$ was determined by using the following equation:

$$t1 = So\text{Sin}\phi, \tag{2}$$

where So = feed and ϕ = principal cutting edge angle

The VMS [4] was obtained by using the following equation:

$$\text{VMS} = 1.74K(\ln\zeta)^n. \tag{3}$$

2.3 *Theory (Preference Selection Method)*

The following steps were selected to determine the preference selection index:

Step 1. Select relevant criteria and alternatives.

Step 2. Form the decision matrix, Y.

$$Y = \left[Y_{ij}\right]_{m\times n} = \begin{vmatrix} y11 & y12 & \cdots & y1n \\ y21 & y22 & \cdots & y2n \\ \vdots & \vdots & \ddots & \vdots \\ ym1 & ym2 & \cdots & ymn \end{vmatrix}, \tag{4}$$

where Y_{ij} = measured value for the ith alternative and the jth criteria, m = number of alternatives, and n is the number of criteria.

Step 3. The following equations are used to form the normalised decision matrix. To maximise (beneficial) criteria:

$$Y_{ij} = \frac{y_{ij}}{y_{ij\,\max}}, \quad i = 1, 2, 3 \ldots m \tag{5}$$

For minimisation (non-beneficial) criteria: $Y_{ij\min}$

$$Y_{ij} = \frac{y_{ij\,\min}}{y_{ij}}, \quad i = 1, 2, 3 \ldots m \tag{6}$$

Step 4. Find the mean values of normalised performances (Eq. 7).

$$N = \frac{1}{N}\Sigma y_{ij} \tag{7}$$

Step 5. Find the variation in preferences by using the following equation:

$$\Phi_j = \Sigma\left(y_{ij} - N\right)^2 \tag{8}$$

Step 6. Find the deviations from the value of preference by using the following equation:

$$\Omega_j = 1 - \Phi_j \tag{9}$$

Step 7. Find the criteria weight by using the following equation:

$$w_j = \frac{\Omega j}{\Sigma\Omega j} \tag{10}$$

Step 8. Find the PSI of alternatives (Eq. 11).

$$\Theta_i = \Sigma y_{ij} w_j \tag{11}$$

Step 9. Determine the rank of alternatives based on the value of the preference selection index. The largest PSI represents the best rank.

Table 4 Properties of workpieces

Workpiece	Strain-hardening exponent (n)	Strength coefficient (K)	Hardness (BHN)
Medium carbon steel	0.55	5250	168
Inconel 718	0.466	2720	204
D2 cold work steel	0.531	3600	175

3 Results and Discussions

Table 4 displays the strain-hardening exponent, strength coefficient, and hardness values. The strain-hardening exponents for MCS and D2 cold work steel were very high (0.55 and 0.531), and for Inconel 718, the value was comparatively lesser (0.466). A very high value of the strain-hardening exponent (> 0.5) indicates that the hardening is due to the higher dislocation density in the material. This reduces the fracture toughness of the materials. The universal tensile tests also revealed that Inconel 718 has higher fracture toughness and MCS and D2 cold work steel have lower fracture toughness (Fig. 1a–c). The fractured surface of the tensile test specimen for Inconel 718 showed massive deformation and breaking in cup and cone formation, whereas for MCS and D2 cold work steels, breaking of the specimens was predominantly in the brittle mode (Fig. 2a–c).

The performance attributes for the analysis were CRC and VMS (Table 5). The normalised decision matrix is shown in Table 6. All steps of the PSI methodology were followed. Subsequently, Inconel 718-C2 combination was identified as the best-ranked combination, followed by D2 (cold work steel)-C2, and MCS-C1 combinations (Table 7).

4 Rank Validation Study

According to the chip macrograph analysis, unfavourable long (side curling and snarled) chips (Fig. 3a) are formed during MCS machining. As a result, the PSI method ranks them third. When machining Inconel 718, the PSI method ranks first when favourable long tubular (not snarled) chips (Fig. 3b) are formed. D2 cold work steel chips are straight, long, and snarled, showing better form (Fig. 3c) than MCS chips. The second rank has been assigned by the PSI method for D2 cold work steel. Thus, the rank assignment by the PSI method has been validated.

The scanning electron microscopy was done for the chip surfaces. Numerous cracks were observed at the MCS chip top surface (Fig. 4a). However, chip formation occurred by successive shear sliding of the deformed grain in the case of machining Inconel 718 (Fig. 4b). Thus, chip formation for Inconel 718 occurred in an ideal mode. The chip top surface of the D2 cold work steel had numerous cracks (Fig. 4c). The SEM study validates the extensive strain-hardening behaviour of MCS and D2

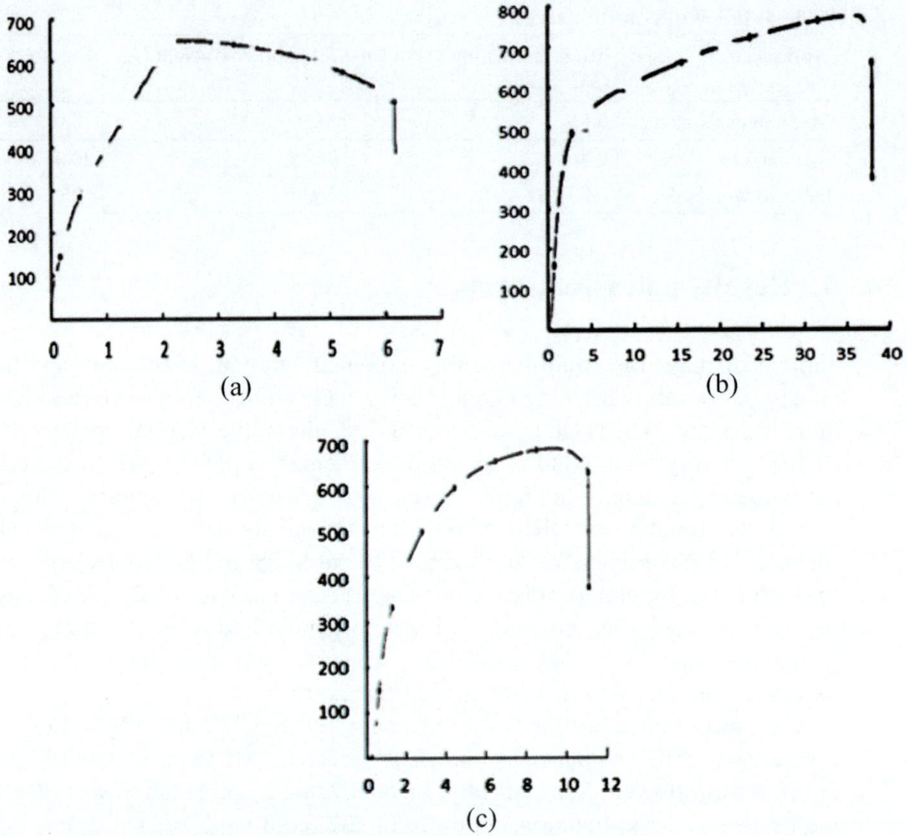

Fig. 1 Stress–strain curves of **a** MCS, **b** Inconel 718, and **c** D2 cold work steel [*X*-axis: Tensile strain (extension, %), *Y*-axis: Tensile stress (MPa)]

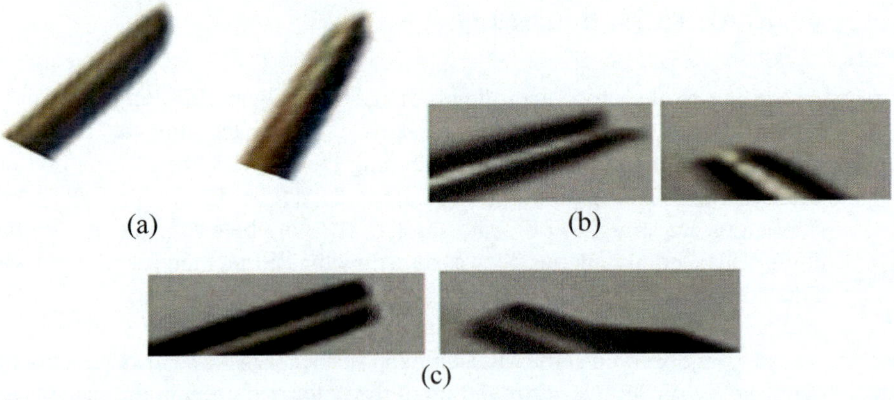

Fig. 2 Fractured specimens of **a** MCS, **b** Inconel 718, and **c** D2 cold work steel

Table 5 Decision matrix

Work-tool combination	CRC	VMS (MPa)
Medium carbon steel (MCS)-C1	3.3174	10,094.74
Inconel 718-C2	1.020	761.011
D2 cold work steel-C2	1.1720	2357.08

Table 6 Normalised decision matrix

CRC	VMS (MPa)
3.3174	10,094.74
1.020	761.011
1.1720	2357.08

The mean values of normalised performances are N = [0.7259 0.4661]
Preferences vary in relation to each criterion = [0.2711 0.4582]
Deviations of the value of the preference are [0.7289, 0.5418, respectively]
Weights of the criteria = [0.5736 0.4264]
Alternative preference selection indices = [0.2085 1 0.6369]

Table 7 Performance selection index values of alternatives and complete ranking

Alternatives	Indexes	Rank
Medium carbon steel (MCS)-C1	0.2085	3
Inconel 718-C2	1	1
D2 cold work steel-C2	0.6369	2

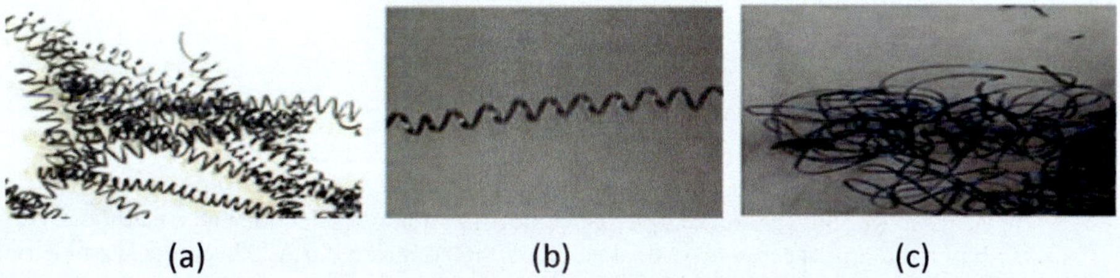

Fig. 3 Chip macrographs **a** MCS, (long side curling and snarled), **b** Inconel 718 (long tubular, not snarled), **c** D2 cold work steel (straight long and snarled)

cold work steels. An ideal chip formation mode for the machining of Inconel 718 also validates the best rank attainment by the PSI method.

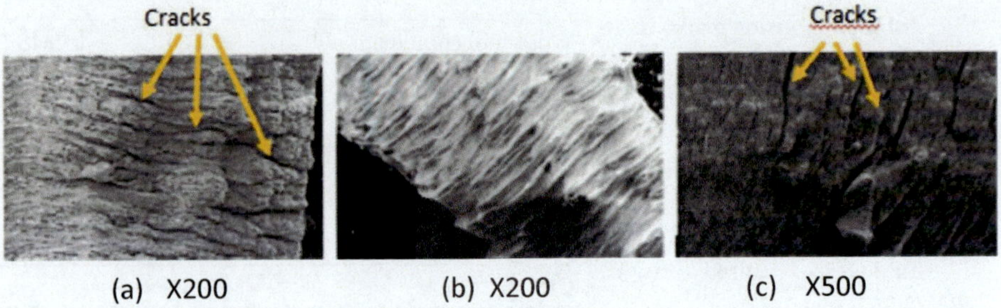

Fig. 4 SEM images of chip (top) surfaces of **a** MCS, **b** Inconel 718, and **c** D2 cold work steel

5 Conclusion

The fracture toughness of the Inconel 718 superalloy was superior in the current list of experimental materials. The strain-hardening exponent for the superalloy was also better in the present list of materials. The machinability of a material depends much on its fracture toughness. The best rank attainment of the Inconel 718 superalloy was attributed to the superior fracture toughness and strain-hardening index of the material. As per the PSI method, the Inconel 718 superalloy attained the best rank, followed by D2 cold work steel and MCS. The qualitative evaluation can validate the rank attainment of the alternatives. The PSI method is very effective for finding the rank of the alternatives.

Acknowledgements Material testing data and SEM images were obtained from IIT Kanpur. The authors are grateful for that.

References

1. Mukesh K, Ashiwani K (2020) Application of preference selection index method in performance based ranking of ceramic particulate (SiO_2/SiC) reinforced AA2024 composite materials. Mater Today: Proc 27:2667–2672
2. Ashiwani K, Mukesh K, Amar P, Pawar MJ, Akhileshwar P, Anil K, Vikas G (2021) Optimization of sliding and mechanical performance Ti/NI metal powder particulate reinforced Al 6061 alloy composite using preference selection index method. Mater Today: Proc 44:4784–4788
3. Kanishka J, Sunil C, Tyagi YK, Hari OM (2018) Characterization of biodegradable composites and application of preference selection index for deciding optimum phase combination. Mater Today: Proc 5:3353–3360
4. Astakhov VP. Tribology of metal cutting. Tribol Interface Eng Ser 52, Elsevier

Analysis of Hole Characteristics in Mechanical Micro-drilling of CFRP-Ti6Al4V Stack Composite

Rahul Kundiya, Raju Pawade, Shivam Yadav, Nayan Sarode, and Jitendra Acharya

Abstract In the present research study, the objective was to estimate three critical output parameters for the mechanical micro-drilling process of CFRP-Ti6Al4V stack composite, which were radial overcut, delamination factor, and hole taper, using response surface methodology (RSM). This study considered spindle speed, feed rate, and MQL flow rate as input parameters and used the central composite design (CCD) method for their experimental design. They conducted 20 experiments and created a mathematical model to determine the relationship between input machining parameters and output parameters. The ANOVA analysis was conducted at a 5% significance level to identify significant coefficients. The study results revealed that spindle speed had a significant impact on all three output parameters. The predicted outcomes from the developed models were consistent with the expected values. The experimental results demonstrated a moderate fit, with a coefficient of determination of 0.7482 for radial overcut, 0.7204 for delamination factor, and 0.8423 for hole taper. These findings have practical implications for optimizing the mechanical micro-drilling process for CFRP-Ti6Al4V stack composite, resulting in better hole quality in micro-drilling applications.

R. Kundiya (✉) · R. Pawade · J. Acharya
Department of Mechanical Engineering, VPM's Maharshi Parshuram College of Engineering, Velneshwar, Maharashtra 415729, India
e-mail: rahulkundiya1999@gmail.com

R. Kundiya · R. Pawade
Department of Mechanical Engineering, Dr. Babasaheb Ambedkar Technological University, Lonere, Maharashtra 402103, India

S. Yadav
Department of Mechanical Engineering, Clemson University, 112A University Village Dr, Central, Clemson, SC 29630-4034, USA

N. Sarode
Department of Mechanical Engineering, NYU Tondon School of Engineering, 6 Metrotech Center Freeman Hal, Brooklyn NY11201, USA

N. Kumar et al. (eds.), *Advances in Materials and Agile Manufacturing*,
Lecture Notes in Mechanical Engineering,
https://doi.org/10.1007/978-981-99-6601-1_5

Nomenclature

MQL	Minimum quantity lubrication
CCD	Central composite design
RSM	Response surface methodology
CFRP	Carbon fiber-reinforced plastic
F_d	Delamination factor
D_{max}	Damage zone diameter
d	Hole diameter
d_t	Tool diameter
d_{jt}	Size of the hole the tool left in the workpiece

1 Introduction

CFRP-Ti6Al4V stack composite is extensively used in the airspace industry for manufacturing of engine cowlings, wing connections, wing panels, and nacelles. CFRP/Ti alloy stack composite of about 0.8 mm to 1.25-mm-thick sheet is being used for cowling the high-performance engine of 757 Dreamliner, Falcon 900 business jet, Airbus A320, and F-16XL aircraft. The purpose of an aircraft's engine cowling is to provide the engine with protection as well as effective airflow by minimizing drag, and airflow is directed through the engine to cool it.

In the aerospace industry, advanced composite materials are gaining popularity owing to their superior mechanical properties and excellent structural capabilities. By combining metal and composites, CFRP/Ti stack composites provide a solution to the challenges faced by metal alloys in terms of fatigue strength and corrosion resistance. As a result, they offer increased durability and reduce the need for repair in low-impact energy scenarios [1–3].

Mechanical micro-drilling makes use of peck drilling cycle where the drill is repeatedly inserted and withdrawn from the hole, which leads to the creation of a hole. The potentiality of the technique may be visible through its ability to manufacture holes having a diameter as small as 50 μm on substances like plastics and polymers as well as metals. The demand for micro-drilling, that is, drilling holes of diameter much less than 1 mm, is gaining significance because of the growth in using miniature and micro-gadgets. The predominant areas of application of micro-drilling are cars, aeronautics, electronics, and clinical fields [4, 5].

It is frequently necessary to drill millions of holes in stacked materials to suit the assembling requirements of many industries. Hybrid composite stacks are known to be tough to cut and provide issues while drilling, as discussed in a review by Xu et al. The distinct characteristics of the stacked pieces, as well as their poor machinability, are the primary causes of this problem. Delamination in the hybrid phase and Ti burrs in the metal phase commonly result in severe hole damage during the material removal process of drilling, which accounts for a significant portion of component rejections. In addition, machining costs rise as a result of quick tool wear and early tool failure. Most rejected parts are assumed to be the result of inefficient cutting parameters, poor drill bits, and inappropriate cutting conditions [1, 6, 7].

There are two cutting sequence options available for the mechanical micro-drilling procedure of the CFRP-Ti6Al4V stack: drilling from CFRP/Ti6Al4V and drilling toTi6Al4V/CFRP. The bimaterial system's machinability can be considerably enhanced by the cutting sequence method employed. A review of the literature revealed that numerous academics think the CFRP-Ti alloy drilling sequence is a practical and effective approach to minimize hole damage while drilling the composite material vertically. The bottom Ti alloy contributes by reducing workpiece dynamics and laminate deflection, which results in less exit delamination damage [1, 2, 6–8].

Landge [9] performed research on the micromechanical drilling of challenging-to-cut materials, focusing on the mechanical micro-drilling of a brass alloy. The researchers looked at a variety of input machining factors, including speed, feed, hole depth, and cutting time. Their results indicated that speed was the most crucial factor affecting the drilling process [9].

A study focused on tool wear and surface quality was conducted on the machining of CFRP-Ti alloy stack using tungsten carbide tools. The study revealed that increasing feed rates led to higher degrees of delamination, burr levels, and surface roughness, while increasing cutting speed resulted in a decrease in these factors. The use of multi-facet stacks led to significantly reduced drill lifetimes because of drill wear coupling [1]. In a separate study, SenthilKumar et al. studied the wear mechanism of tools with different geometrical characteristics when machining of CFRP/Ti6Al4 V stacks. The findings suggested that drills with a 130° point angle and smaller chisel edge thickness exhibited lower levels of wear and better chip ejection capability [10].

Cong et al. investigated the application of ultrasonic vibration assistance (UVA) during the mechanical micro-drilling of CFRP-Ti alloy stacks. UVA drilling greatly enhanced the machinability of these composite materials, according to the researchers [11].

Shunmugesh and Panneerselvam [12] performed research on the micromechanical drilling of CFRP and utilized the Taguchi methodology to optimize the machining process. The study evaluated input parameters such as drill diameter, spindle speed, and feed rate and developed a mathematical model for circularity error and cylindricity error. The study determined that the optimal conditions for minimizing circularity and cylindricity errors were the maximum spindle speed and the minimum feed rate [12].

A useful technique for comprehending how input variables and their interactions impact response variables in manufacturing processes is the response surface methodology (RSM). RSM creates models that examine the intricate correlations between input variables using statistical approaches like central composite design. The model's performance is then assessed using ANOVA checks. RSM also makes use of 3D response graphs to show how input variables influence results. RSM has been widely used by researchers to evaluate the effectiveness of industrial processes [13–16].

After reviewing the literature, it was discovered that there have been no previous research studies conducted on micromechanical drilling of the CFRP-Ti6Al4V composite material stack. Therefore, this particular study employed RSM to create a mathematical model for the output responses, which included delamination factor, hole taper, and radial overcut. The study considered input parameters such as spindle speed, feed rate, and minimum quantity lubrication (MQL) flow rate. To evaluate the effectiveness of the model, ANOVA tests were performed, and 3D response graphs were utilized to analyze the impact of machining parameters on output responses.

2 Materials, Methods, and Processes

The experiments were carried out on SMD10B CNC micro-drilling machine manufactured by Interface design associates private limited, India. Figure 1 shows theSMD10B CNC micro-drilling machine. The aerostatic spindle is provided to vary the speed between 6001 and 60,000 rpm on the machine. Brown translucent emulsion oil was used as lubrication oil. For the experiments, a stack composite of CFRP and Ti6Al4V grade 5 was used as the workpiece material. Table 1 depicts the mechanical properties of Ti6Al4V and CFRP material. Table 2 depicts the specification of CNC micromachining station.

Araldite AW134 Epoxy resin was used for stacking CFRP and Ti6Al4V. The tool material used for the present experiment was T5401 grade—TiN-coated tungsten carbide (coating thickness 0.2–0.5 microns). For this study, a micro-drill that has a diameter of 0.4 mm and a point angle of 140° was chosen.

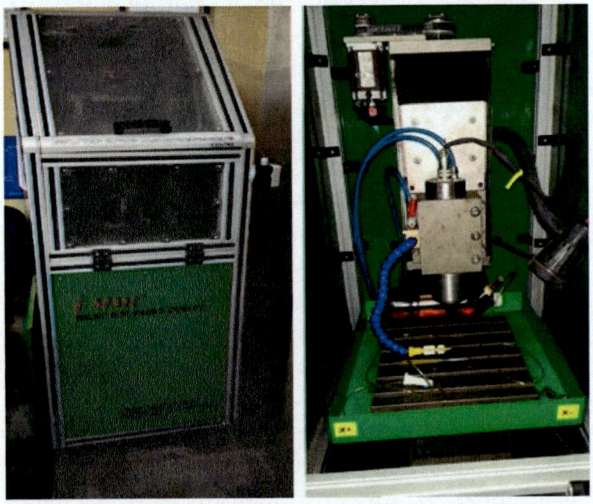

Fig. 1 CNC micromachining station for micro-drilling

Table 1 Mechanical properties of CFRP and Ti6Al4V

Properties	CFRP	Ti6Al4V
Density, g/cm^3	1.9	4.42
Young's modulus, GPa	250	113.8
Shear modulus, GPa	110	44
Uniform elongation, %	1.9	14
Poisson's ratio	0.33	0.342
Hardness HV		349
Yield strength (MPa)		880
Tensile strength (MPa)		950

Table 2 Specifications of the CNC machine model interface, SMD10B

XY axis travel	100 mm × 100 mm
Support	LM guideways
Accuracy	5 micron/25mm
Max feed	200 mm/min
Spindle speed	200 watts, 60,000 rpm, 2 microns runout, ER11 collet

Figure 2 illustrates the use of ALICONA Infinite Focus, an optical microscope, to measure the diameter of the holes formed in the work material. To calculate the radial overcut, the approach described in Veenaraja et al. [17] was adopted. This involved determining half the difference between the hole diameter and the tool diameter, as indicated in Eq. 1.

$$\text{Radial overcut} = \frac{d_{jt} - d_{\mathrm{t}}}{2}. \tag{1}$$

The process of drilling is a commonly used technique for the assembly of various structures or components. However, the drilling of composite laminates is often associated with a type of damage known as delamination. Compared to the drilling of metals, drilling composites poses a greater risk of delamination, which is considered a major issue in the field.

The delamination factor was determined by using Eq. 2 as [18]:

$$F_{\mathrm{d}} = \frac{D_{\mathrm{max}}}{d}. \tag{2}$$

Hole taper is one of the inherent manufacturing problems associated with micro-drilling. A taper is defined as the uniform change of diameter on its axis. The taper of holes determines the quality of micro-holes. In addition to the size requirements, the micro-holes require no taper or minimum taper.

The hole taper was determined by using Eq. 3 as [19]:

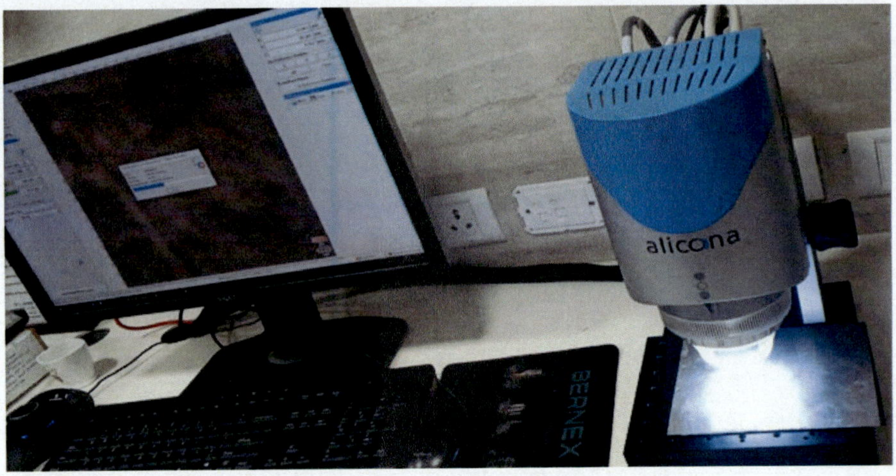

Fig. 2 Measuring diameter of machined micro-holes on work material by ALICONA optical microscope

$$\text{Taper} = \frac{D - d}{L}, \tag{3}$$

where $D =$ Large diameter, $d =$ Large diameter and $L =$ Length.

3 Experimental Design and Parameter Selection

RSM is a practical method for creating, enhancing, and maximizing processes. It entails examining how intricate interconnections between input factors affect how response variables behave. Typically, Eq. 4 performance describes the response surface [20]:

$$Y = \beta_0 + \sum_{i=1}^{s} \beta_i \chi_i + \sum_{i=1}^{s} \beta_i \chi_i + \sum_{i=1}^{s} \beta_{ii} \chi_i^2 + \sum_{i=j}^{s} \beta_{ij} \chi_i \chi_j + \varepsilon. \tag{4}$$

In the present study, RSM was adopted for developing a mathematical model using statistical design of experiments approaches, specifically the central composite design. To evaluate the models performance, the ANOVA test was conducted. The experimental design was implemented using the CCD approach, and the Design Expert 13.0 statistical program was utilized to determine the regression model coefficients from the experimental data. The machining input parameters, including spindle speed, feed rate, and minimum quantity lubrication (MQL) flow rate, were selected based on the literature on mechanical micro-drilling research and the operational

Table 3 Selected machining parameters and their levels

Name	Units	Lowest	Low	Center	High	Highest
Spindle speed	rpm	26,591	30,000	35,000	40,000	43,408
Feed rate	mm/min	1.03	1.2	1.45	1.7	1.87
MQL flow rate	ml/hr	99	150	225	300	351

characteristics of the chosen machine. Table 3 presents with the chosen machining parameters and their respective levels.

4 Results and Discussion

According to the input data in Table 4, 20 experimental runs for the CCD were carried out. The machined holes diameter on work materials can be measured using an optical microscope (ALICONA), as shown in Fig. 2. The output response which include radial overcut, delamination factor, and hole taper is then determined for each run using Eqs. (1), (2), and (3) are summarized in Table 4 [14–16].

5 Analysis of Hole Characteristics

This study aimed to examine the hole characteristics, such as radial overcut, delamination factor, and hole taper mechanism, during mechanical micro-drilling. With the use of the Design Expert program, a mathematical model was constructed using experimental data to determine the link between machining settings and radial overcut. To evaluate the statistical significance of the quadratic model and investigate the effects of machining factors on output responses such as radial overcut, delamination, and taper, ANOVA analysis was performed. The results revealed that the model was remarkably significant and could be utilized to predict the impact of various machining factors on the output responses.

The model F-value of 3.30 suggests that the model is noteworthy. In Table 5, model terms with values of "Prob > F" are deemed statistically significant if they are smaller than 0.05 (95% confidence). There is no discernible difference between the lack of fit and the pure error, according to the lack of fit F-value of 0.8510. Furthermore, the model appears to account for roughly 74.82% of the variance in the ROC, according to the derived R-squared value of 0.7482. The signal in this model is excellent, as indicated by the adequate precision score of 5.5638, which is higher than 4.

Table 4 CCD design layout and experimental results

Run	Factor 1	Factor 2	Factor 3	Response 1	Response 2	Response 3
	A: Spindle speed RPM	B: Feed rate mm/min	C: MQL flow rate ml/hr	Radial overcut	Delamination factor	Hole taper
1	30,000	1.2	150	3.45	1.0064	1.254
2	40,000	1.2	150	28.24	1.0429	1.025
3	30,000	1.7	150	12.6	1.0611	0.985
4	40,000	1.7	150	15.7	1.0101	0.4596
5	30,000	1.2	300	9.21	1.0381	0.595
6	40,000	1.2	300	11.13	1.1374	0.215
7	30,000	1.7	300	17.6	1.0083	0.428
8	40,000	1.7	300	20.8	1.0623	0.1104
9	26,591	1.45	225	3.35	1.0094	2.13
10	43,408	1.45	225	32.25	1.0761	0.73
11	35,000	1.02955	225	25.51	1.0591	0.14
12	35,000	1.87045	225	20.11	1.0095	0.1263
13	35,000	1.45	98.8655	2.36	1.0421	0.456
14	35,000	1.45	351.134	7.21	1.0936	0.32
15	35,000	1.45	225	5.12	1.1067	0.743
16	35,000	1.45	225	11.56	1.0079	0.889
17	35,000	1.45	225	8.35	1.0398	0.6032
18	35,000	1.45	225	0.11	1.0317	0.95
19	35,000	1.45	225	8.35	1.0525	0.4
20	35,000	1.45	225	19.56	1.0031	0.689

Backward elimination removes non-significant terms from the fitted quadratic model before it is adjusted for ROC. The following is how the final radial overcut quadratic model is arrived at (Fig. 3):

$$\text{ROC} = 8.86 + 10.05A - 7.21AB + 13.54B^2. \tag{5}$$

When the data's underlying assumptions about normality are tested, it is evident that every point on the normal plot is nearly straight. The expected values for the radial overcut match the experimental results. Similarly, the expected values for delamination factor and hole taper match the experimental result.

The quadratic model for the delamination factor was established as follows:

$$F_{\text{d}} = \left(9.65433 \times 10^{-6}\right)A + 0.733408B + \left(5.5933 \times 10^{-3}\right)AC - 0.043845B^2. \tag{6}$$

Table 5 Analysis of variance for radial overcut model

Source	Sum of squares	df	Mean square	F-value	p-value	
Model	1170.52	9	130.06	3.30	0.0383	**Significant**
A-Spindle speed	487.63	1	487.63	12.38	0.0056	
B-Feed rate	2.29	1	2.29	0.0581	0.8144	
C-MQL flow rate	3.50	1	3.50	0.0888	0.7719	
AB	52.07	1	52.07	1.32	0.0421	
AC	64.81	1	64.81	1.64	0.2286	
BC	57.51	1	57.51	1.46	0.2548	
A^2	131.91	1	131.91	3.35	0.0972	
B^2	331.59	1	331.59	8.42	0.0158	
C^2	35.80	1	35.80	0.9086	0.3630	
Residual	393.99	10	39.40			
Lack of fit	181.14	5	36.23	0.8510	0.5681	**Not significant**
Pure error	212.85	5	42.57			
Cor total	1564.51	19				
R^2	0.7482					
Adjusted R^2	0.5215					
Predicted R^2	0.5124					
Adequate precision	5.5638					

The quadratic model for the hole taper was established as follows:

$$\text{Hole taper} = -0.4987A - 0.3207B + 0.7100A^2 - 0.5855B^2. \quad (7)$$

5.1 *Effect of Machining Parameters on the Output Parameter*

Radial overcut is an intrinsic feature of the hole produced through micro-drilling and cannot be avoided, although compensations can be made through tool design. To enhance the precision of the mechanical micro-drilling process, it is crucial to minimize radial overcut. As a result, the parameters influencing radial overcut must be identified.

Figure 4a, b illustrates the relationship between spindle speed, feed rate, MQL flow rate, and radial overcut. In Fig. 4a, it can be shown that the optimal range of spindle

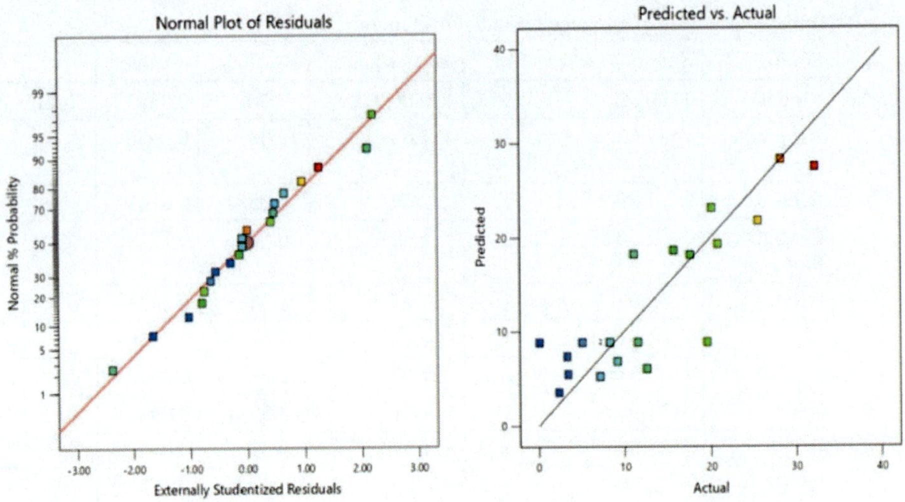

(a) (b)

Fig. 3 **a** Normal probability plot for radial overcut, **b** comparison with predicted radial overcut with experimental data

speed and feed rate to minimize radial overcut is achieved when both parameters are set to intermediate levels. The results suggest that a moderate feed rate can be advantageous in reducing radial overcut. In contrast, Fig. 4b presents the estimated surface response for radial overcut concerning spindle speed and MQL flow rate. The graph indicates that the minimum radial overcut is attained at a minimum MQL flow rate and a spindle speed that is close to the middle level. Additionally, the results reveal that higher spindle speeds lead to higher radial overcut values.

The effect of feed rate and flow rate on radial overcut is shown in Fig. 4c. The graph highlights that the lowest radial overcut is seen at the intermediate level of MQL flow rate and the middle level of feed rate. Additionally, it is noteworthy that a moderate spindle speed level is more favorable for achieving lower radial overcut. Furthermore, the findings indicate that the MQL flow rate has a more significant effect on radial overcut than the feed rate and spindle speed.

According to the trends depicted in the graphs, there is a direct correlation between the spindle speed and the centrifugal force that the lubrication fluid, which contains debris particles, experiences in the gap between the tool and workpiece. Therefore, a slight increase in spindle speed can assist in reducing the radial overcut. However, a higher feed rate coupled with a rapid increase in spindle speed can result in increased friction between the tool and workpiece, causing higher heat generation and wear. This, in turn, leads to larger radial overcut.

Delamination is one of the inherent manufacturing problems associated with the micro-drilling of composite material. Delamination should be minimum to achieve greater accuracy in the micro-drilling process. The graph presented in Fig. 4d demonstrates the correlation between spindle speed, feed rate, and delamination factor. The

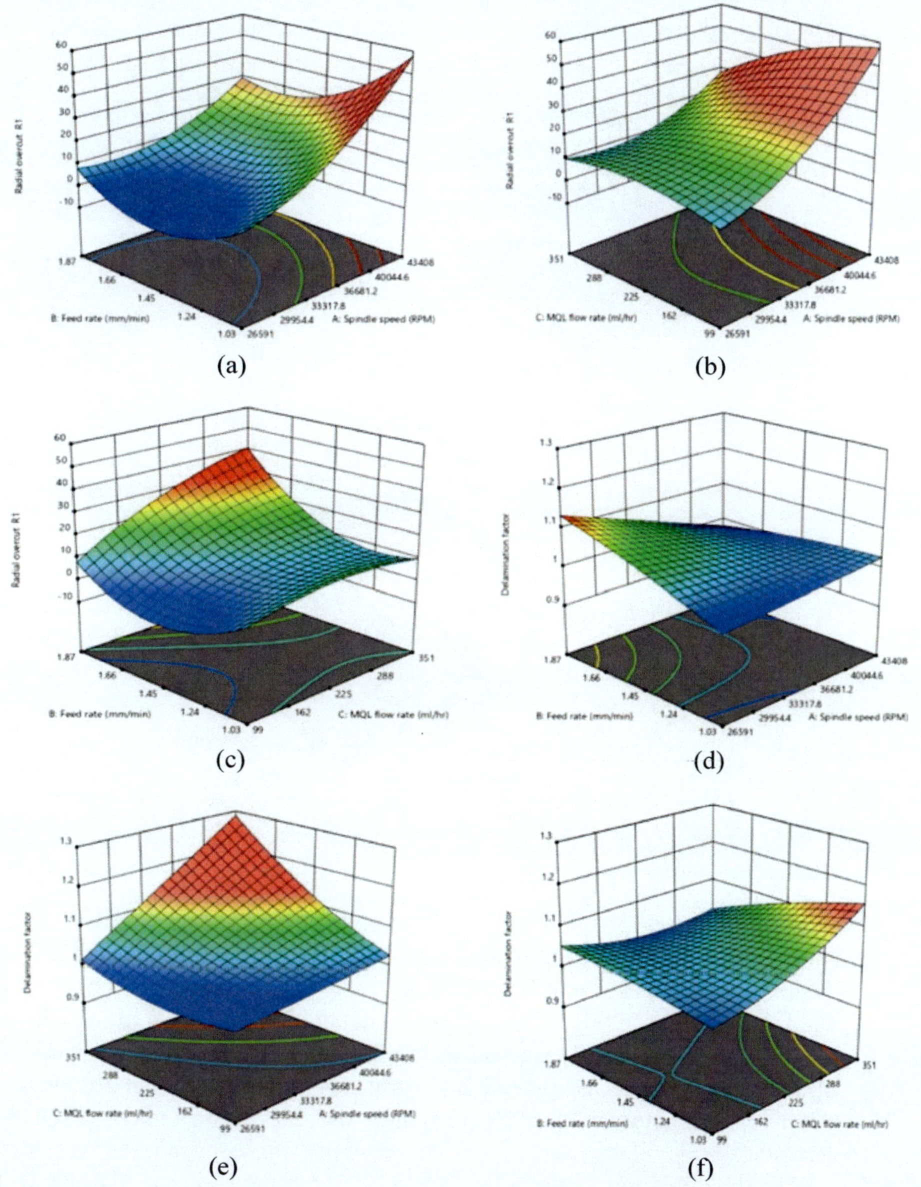

Fig. 4 Response surface plot for **a** radial overcut versus spindle speed and feed rate, **b** radial overcut versus spindle speed and MQL flow rate, **c** radial overcut versus feed rate and MQL flow rate, **d** delamination factor versus spindle speed and feed rate **e** delamination factor versus spindle speed and flow rate, **f** delamination factor versus MQL flow rate and feed rate, **g** hole taper versus spindle speed and feed rate, **h** hole taper versus spindle speed and MQL flow rate, **i** hole taper versus MQL flow rate and feed rate

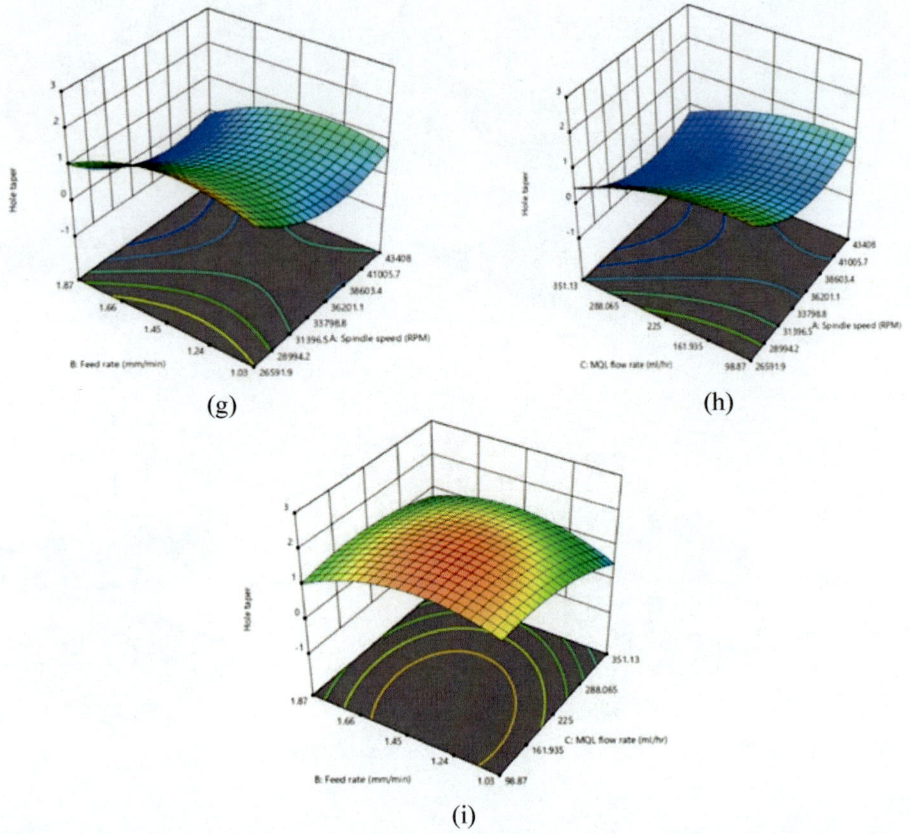

Fig. 4 (continued)

optimum combination of spindle speed and feed rate for minimizing the delamination factor is observed to be at intermediate levels. Additionally, it is observed that the feed rate has a more significant effect on delamination compared to spindle speed. In Fig. 4e, the estimated response surface for the delamination factor with respect to spindle speed and MQL flow rate is illustrated. The findings reveal that the maximum delamination factor occurs at a high spindle speed level and a middle MQL flow rate level.

The delamination factor is determined by both the feed rate and the MQL flow rate, as shown in Fig. 4f. According to the graph, the lowest delamination factor is obtained at a medium feed rate and when the MQL flow rate is similar to medium. Moreover, the results show that an intermediate spindle speed resulted in a decreased delamination factor. In terms of machining parameters, the influence of spindle speed and feed rate on the delamination factor is greater than the effect of MQL flow rate on the same factor.

Based on the trend observed in the graph, it can be concluded that an rise in spindle speed and drill diameter can minimize the delamination factor and circularity errors.

Additionally, it can be noted that the feed rate and drill diameter have a significant impact on delamination damage, which becomes more pronounced with the increment of tool wear. The rise in spindle speed minimizes delamination tendency; therefore, it reduces circularity error mainly affected due to reduction in vibration effect of drill bit and machine.

Reduced feed rate and increased spindle speed have been identified as effective measures in reducing delamination in CFRP, as they lead to a decrease in the concentration of thrust force on the material. When using smaller drill bits, the dissipation of heat is slower compared to larger drill bits, which could result in the development of very high temperatures at actual spindle speeds. This could cause micro-fractures and increase tool wear.

Micro-drilling often results in tapered holes, which can negatively affect the accuracy of the process. Figure 4g illustrates the impact of spindle speed and feed rate on hole taper. To reduce taper in micro-drilling, it is necessary to optimize the spindle speed and feed rate. According to the experimental results, the best settings for achieving the least amount of taper are a medium spindle speed and a higher feed rate. The lowest taper is observed at a higher feed rate and near the middle range of spindle speed, as indicated by the predicted response surface for taper in Fig. 4h. It is important to note that maximum hole taper occurs at lower spindle speeds.

Figure 4i depicts the influence of spindle speed and feed rate on taper. According to the graph, the minimum taper is obtained by using a higher level of MQL flow rate and a middle level of feed rate. The graph's trend indicates that spindle speed is an important factor in determining machining quality. A rougher surface of the hole results from a higher feed rate and spindle speed. Furthermore, spindle speed increases drill stiffness, resulting in better circular holes with minimal taper. This is because the drill bit is self-piloting and has more stability at higher speeds.

6 Conclusion

Response surface approach was utilized in this work to estimate the radial overcut in micromechanical drilling of CFRP-Ti6Al4V stack composite material. The central composite design (CCD) approach was used to run 20 tests with three input parameters at five levels. Based on the experimental data, RSM was used to build mathematical models for radial overcut, delamination factor, and hole taper. The ANOVA findings show that spindle speed is very important for the radial overcut reaction, although feed rate and MQL flow rate are not. The delamination factor was significantly significant for spindle speed and feed rate, but not for MQL flow rate. For hole taper, spindle speed and MQL flow rate were extremely relevant, although feed rate was not. The predicted values were in close agreement with the actual data, with coefficients of determination of 0.7482 for radial overcut, 0.7204 for delamination factor, and 0.8423 for hole taper. The lowest radial overcut was produced at the medium levels of spindle speed and feed rate, and the intermediate feed rate was the best for minimizing the delamination factor, according to the response surface plots.

The lower taper was seen at higher flow rates and spindle speeds around the middle values. Overall, the findings show that response surface approach is an excellent way for modeling the input machining parameters of the mechanical micro-drilling process for CFRP-Ti6Al4V stack composites. It also implies that in future research, this strategy may be used to construct models for additional reactions, such as MRR and SR.

References

1. Xu J, el Mansori M (2016) Experimental study on drilling mechanisms and strategies of hybrid CFRP/Ti stacks. Compos Struct 157:461–482. https://doi.org/10.1016/j.compstruct.2016.07.025
2. Isbilir O, Ghassemieh E (2013) Comparative study of tool life and hole quality in drilling of CFRP/titanium stack using coated carbide drill. Mach Sci Technol 17(3):380–409. https://doi.org/10.1080/10910344.2013.806098
3. Shyha I, Soo SL, Aspinwall DK, Bradley S, Dawson S, Pretorius CJ (2010) Drilling of titanium/CFRP/aluminium stacks. Key Eng Mater 447(448):624–633. https://doi.org/10.4028/www.scientific.net/KEM.447-448.624
4. Hasan M, Zhao J, Jiang Z (2017) A review of modern advancements in micro drilling techniques [Online]. Available: https://ro.uow.edu.au/eispapers1/758
5. Heinemann R, Hinduja S, Barrow G, Petuelli G (2006) The performance of small diameter twist drills in deep-hole drilling. J Manuf Sci Eng 128(4):884–892. https://doi.org/10.1115/1.2335859
6. Pecat O, Brinksmeier E (2014) Tool wear analyses in low frequency vibration assisted drilling of CFRP/Ti6Al4V stack material. Procedia CIRP 14:142–147
7. Lin YC, Tsao CC, Hsu CY, Hung SK, Wen DC (2012) Evaluation of the characteristics of the microelectrical discharge machining process using response surface methodology based on the central composite design. Int J Adv Manuf Technol 62(9–12):1013–1021. https://doi.org/10.1007/s00170-011-3745-0
8. Natarajan U, Periyanan PR, Yang SH (2011) Multiple-response optimization for micro-endmilling process using response surface methodology. Int J Adv Manuf Technol 56(1–4):177–185. https://doi.org/10.1007/s00170-011-3156-2
9. Landge RR (2017) Analysis of micro-drilling process using response surface methodology [Online]. Available: http://www.ripublication.com
10. SenthilKumar M, Prabukarthi A, Krishnaraj V (2013) Study on tool wear and chip formation during drilling carbon fibre reinforced polymer (CFRP)/titanium alloy (Ti6Al4V) stacks. Procedia Engg 64:582–592
11. Prabukarthi A, Senthilkumar M, Krishnaraj V (2016) Study on drilling of CFRP/Ti6Al4V stack with modified twist drills using acoustic emission technique. Steel Compos Struct 21:573–588
12. Shunmugesh K, Panneerselvam K (2018) Multi-performance optimization of micro-drilling using Taguchi technique based on membership function
13. Djoudi W, Aissani-Benissad F, Bourouina-Bacha S (2007) Optimization of copper cementation process by iron using central composite design experiments. Chem Eng J 133(1–3):1–6. https://doi.org/10.1016/j.cej.2007.01.033
14. Sohani MS, Gaitonde VN, Siddeswarappa B, Deshpande AS (2009) Investigations into the effect of tool shapes with size factor consideration in sink electrical discharge machining (EDM) process. Int J Adv Manuf Technol 45(11–12):1131–1145. https://doi.org/10.1007/s00170-009-2044-5
15. Tsao CC (2008) Comparison between response surface methodology and radial basis function network for core-center drill in drilling composite materials. Int J Adv Manuf Technol 37(11–12):1061–1068. https://doi.org/10.1007/s00170-007-1057-1

16. Gopalakannan S, Senthilvelan T (2013) EDM of cast Al/SiC metal matrix nanocomposites by applying response surface method. Int J Adv Manuf Technol 67(1–4):485–493. https://doi.org/10.1007/s00170-012-4499-z
17. Veenaraja D, Muthukumar V, Venkatasamy R, Dharmendhirakumar M, Sureshbabu A, Senthilkumar N (2013) Impact of machining parameters on the EDM process responses: a Taguchi approach for Al-SiC MMC. In: 4th Nirma university international conference on engineering, 2013
18. Jawalkar C, Kant S. Analysis of delamination factor in drilling using Taguchi and regression method
19. Chen Q, Wang HJ, Lin DT, Zuo F, Zhao ZX, Lin HT (2018) Characterization of hole taper in laser drilling of silicon nitride ceramic under water. Ceram Int 44(11):13449–13452. https://doi.org/10.1016/j.ceramint.2018.04.173
20. Habib SS (2009) Study of the parameters in electrical discharge machining through response surface methodology approach. Appl Math Model 33(12):4397–4407. https://doi.org/10.1016/j.apm.2009.03.021

Redefining Manufacturing: The Design Thinking Advantage

Jyoti Jinagal Karloopia and Rajat Agrawal

Abstract Many business leaders continue to prioritise, yet few are satisfied with their businesses' innovation skills. Many businesses experiment with innovation approaches brought in from the outside in their desire to become more innovative. In recent years, one of these sought-after innovation methodologies has been design thinking which is a mindset and a set of tools for innovation and development. Design thinking has created spectacular achievements in creative organisations like design consultancies and service firms but it has shown to be significantly more difficult to execute in large, established firms, especially if the organisational culture opposes the ideals that underpin design thinking. We demonstrate how a large, manufacturer of farm equipment's enterprises over a period of five years morphed the design thinking methodology to allow it to take root in many sectors of the organisation and become an intrinsic element of the firm's innovation activities over time by including design thinking into their strategic plan, leadership programmes and into day-to-day activity. This paper also throws some light on other approaches which can be used for adopting design thinking in large organisations.

Keywords Design thinking · Manufacturing firm · Innovation · Leadership · Strategy

1 Introduction

Large manufacturing firm reports that being innovative is difficult [7], and academics argue that this is due to a strong focus on efficiency and short-term profits [10]. Manufacturers have hitherto focused solely on product features, with little regard for how the product was really utilised or what behaviours end customers displayed. To solve this issue, many businesses are attempting to develop new ways of working

J. J. Karloopia (✉) · R. Agrawal
Department of Management Studies, Indian Institute of Technology Roorkee, Roorkee, Uttarakhand 247667, India
e-mail: jjkarloopia@bm.iitr.ac.in

N. Kumar et al. (eds.), *Advances in Materials and Agile Manufacturing*,
Lecture Notes in Mechanical Engineering,
https://doi.org/10.1007/978-981-99-6601-1_6

to boost their innovativeness [8]. Design thinking (DT) has gained traction in the business in recent years. The concept of DT has recently arisen in management literature, illustrating how any company can benefit from designers' work [3, 12]. One of the most essential aspects of the concept is that it implies that anyone can learn to use a design approach to solve any innovative problem [13]. The concept of DT is gaining popularity, although it remains vague. Mainly, DT is being implemented in an organisation through various programmes and consultancy projects. As a result, it is being used in many industries.

With large organisations' current focus on effectiveness, there is a natural requirement for evaluating concepts and their influence on the organisation. However, implementing new concepts like DT is often difficult [19], as it puts existing ways of working to the test and creates ambiguity and uncertainty about the new practice and its impact. Rauth et al. [15] found that perceived values associated with the application of DT in big companies go beyond project production and include the development of long-term innovation capacities another difficult to measure feature. Lockwood [11] argues that there is no better time than now for creative problem-solving and new ways of thinking and we require new, transformative corporate strategies based on human needs rather than financial analyses, and that design is the key to achieving all of this because it drives innovation. As a result, DT implementation in a large business will always be questioned as in which way design thinking can be implemented, and there will be a slew of challenges to overcome. As a result, this paper aims to fill in some of the gaps in our understanding by presenting how design thinking can be implemented in an organisation through project-based approach. Based on the information extracted from the firm, the project-based approach is discussed in detail, and three new ways of implementation of design thinking are also discussed.

2 Literature Review

In the manufacturing industry, innovation is viewed as a critical component of increasing productivity. This is especially significant in high-cost operating contexts when non-price-based competition is required. Non-priced-based competition is made up of several intricate tactics that ensure inimitable performance [16]. During the first decade of the twenty-first century, design thinking arose as a method for promoting radical and gradual innovation in businesses seeking to develop new goods and services [14]. To achieve non-price-based competition, the industrial sector is increasingly turning to innovation to ensure productivity growth, particularly in high-cost operational situations. The development of a deep understanding of the user is at the heart of design thinking, allowing you to empathise with their environment and individual difficulties. When producing new products, design thinking is the practice of focusing on the end user's experience and matching people's demands. As factories become more networked, adopting this method in a production facility is a means to obtain a competitive advantage.

Design thinking, according to Lockwood [11], is a human-centred innovation approach that stresses observation, collaboration, fast learning, idea visualisation, rapid concept prototyping, and business strategy. Design thinking, according to Brown [4], is a powerful approach to innovation that is not only effective but also widely accessible, able to be integrated into many sectors of business and society, and valuable for generating breakthrough ideas that can be adopted and have an impact. Design is essentially a creative activity, whereas management is a decision-making process [9]. Instead of choosing between well-defined options, design thinkers' goal is to meet specific human needs. They deal with problems that are difficult to define and improve the efficiency of businesses. Furthermore, DT is sometimes portrayed as a magical solution that will enable innovation assist organisations in developing breakthrough ideas [3] and allow firms to strike the correct balance between exploration and exploitation [13].

Design thinking uses a method of prototyping to reduce the risk in a business model concept by testing it with the marketplace; it allows for the creative development of an idea [18]. Design thinking when integrated into an organisation, it can provide a competitive edge and develop innovation capabilities of employees. Dorst and Cross [6] delve deeper into the iterative nature of design thinking. It is an iterative process "marked by trial-and-error learning that examines a variety of possible solutions with end-users and other project stakeholders" [2]. Prototyping is viewed as a technique to experiment and develop concepts, rather than to finalise them, just as empathy is a way to be user-centred [17].

3 Research Methodology

A case-based research methodology was adopted for this study. A in-depth interview was done with one employee of this firm. He was a top-level management employee who handled the operational activities of the organisation. Given the paucity of empirical knowledge about DT, the study took a case-based, exploratory approach to data collection [1]. The goal is to find the existing way through which design thinking has been implemented in their organisation and they also discussed new ways in which design thinking can be implemented in an organisation. We identified the firm by seeing their official website where they have mentioned that they use design thinking in their organisation. Most of the data was collected through in-depth interviews, and some information was accessed through the internal documentation available. The interview was focused on what is design thinking to them, how they are using it, why there was a need to adopt design thinking, and how they have expanded it in their organisation. Table 1 contains the content of the case study protocol.

Table 1 Case study protocol content

S. No.	Instrument	Description
1	Organisation	Farm equipment's manufacturing firm
2	Permission	The employee was ready to give interview
3	Objective	To find out the key takeaways of using project-based approach as one of the ways for implementation of design thinking in a firm
4	Unit of analysis	Design thinking led projects
5	Source of data and Analysis	Interview
6	Construct validity	Through the employee of the organisation
7	Key question	Can project-based approach is a successful approach and how further design thinking is diffused in the firm

4 Findings

4.1 About the Firm

This manufacturing firm is a 40-year-old organisation which deals in manufacturing of farm equipment's. They have their presence in 100 countries. They have been contributing to agriculture by delivering new implements at the most reasonable costs. They have established a significant presence in both the domestic and international markets because of our relentless pursuit of quality in all aspects of the business. Their cutting-edge manufacturing facilities are outfitted with cutting-edge gear capable of completing large-scale orders with precision. They manufactured creative implements of international standards using the highest quality raw materials and sustainable technical ideas.

4.2 Triggering Point

To boost innovation culture in organisation, the top-level management of the organisation decided to adopt new approaches of innovation. Design thinking which is defined as a process that allows a company to analyse and assess radical new propositions from a variety of angles, typically encompassing user needs, business objectives, and technological demands [5] was selected as one of the approaches as they have found this approach to be very successful in other industries. Now, the main question that aroused among the top-level executive is to how to implement design thinking into the organisation and what all challenges they will be facing. Even though innovation was described as "essential" to the organisation, there was no structured approach as to how to achieve it. After much consideration, a project-based approach was finalised to implement design thinking into the organisation.

4.3 Implementation of Design Thinking Through Project-Based Approach

As per the respondents, design thinking implementation started with the approval initiatives of top management, and resources for the same were also made available by the top management and decision-makers of the organisation. Human resource department played a very important role in adopting DT in organisation, and lastly, a proper feedback mechanism was created to know how this approach is working in the organisation. All four important pillars are discussed in elaborated way.

4.3.1 Top-Level Management/ Leadership

As we know, any new initiative cannot be successful without its alignment with the vision and mission of the organisation; our findings are in line with DT as well. Our interviewers told us that after the decision of implementing design thinking into the organisation through project-based approach, a detailed cost–benefit analysis (CBA) report was submitted to the top-level management which tells about what all cost will be incurred while implementing it and what are the benefits of using design thinking. With top-level approval and taking into consideration, their say a detailed path as to how this approach will be implemented in the organisation was done. Project-based approach was selected as it will save time of the employees as they will learn design thinking while performing the real task.

4.3.2 Availability of Resources

Implementing any new thing requires resources. Resources in terms of money, man, machine, time, and many more things are required for any new thing to start. In the firm, a space was exclusively made for carrying out design thinking led projects. The design thinking room was made different from wall colour of the organisation as to make the look of the space more attractive towards the employees to create an eagerness to know what it is all about. A budget of 15 lakhs was set for one quarter. Employees who will be working on this project were given a brief training of design thinking and made them understand its application and how to use it. A design intern was hired for this project who will guide the employees as how to use design thinking, how much to use, and when to stop using it.

4.3.3 Human Resource Initiatives

As one of the interviewers said, "Design thinking is not only a methodology of innovation to get adopted in an organisation, but also the man behind the method who will be performing it". Design thinking culture development was made a part of HR manual. The HR department made the structure of training which is needed to be given to the employee (target audience) and then they identify the knowledge workers who require these trainings. They developed reward and recognition policies for design led projects which is beneficial to boost its implementation. The finding suggested developing a feedback mechanism after evaluating the project. Awareness creation about the program using multiple approaches such as mailers, word of mouth or a hybrid must also be created. Identification of resource persons for in-house training or, in the case of outsourced training, identification of vendors is also very important; the role of the human resource team is essential in designing, starting, and managing the entire initiative.

4.3.4 Feedback Mechanism

For any programme to get success requires a proper feedback mechanism as to get time to time updates about how the programme is benefitting the employees and the organisation. The HR Head described that they created a proper feedback analysis form which contains questions like:

- What was their (employees) expectation before starting out this project?
- What is the benefit of using design thinking in this project?
- What difference they felt after using design thinking as their approach for developing innovative product?
- What are the new capabilities you have learned while doing this project with design thinking-based approach?
- Do you want this approach to be used further in other projects of the organisation?

All the feedback reports were then submitted to the top-level management for further analysis. The project head also submitted a report to the top-level management which contains all the details as how the project was done, what are the challenges they faced, outcome of the project, and what support they need from the top-level management for future projects (Fig. 1).

4.4 Design Thinking-Based Projects

4.4.1 Project 1 (Case 1)

The organisation thought of bringing a new product called Rotary triller. Therefore, for this innovative product, they tried of using design thinking as an approach for

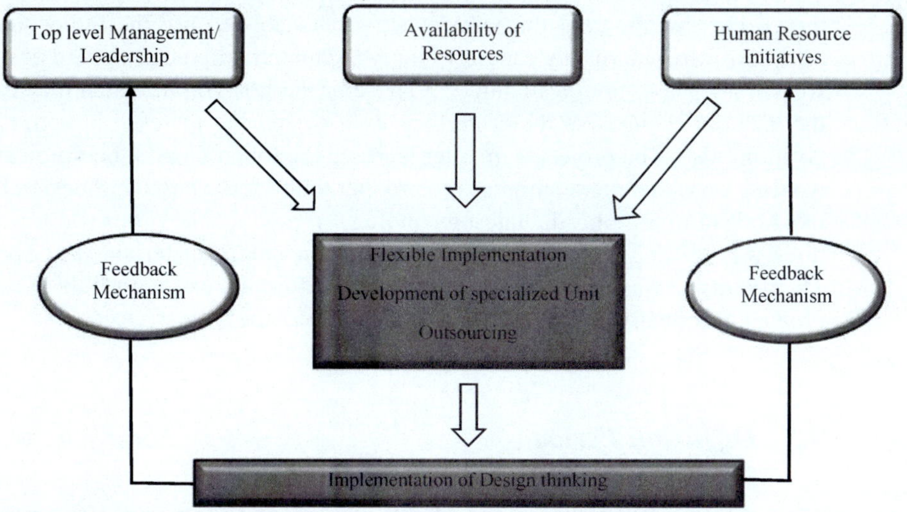

Fig. 1 Implementation of design thinking approach in an organisation

innovation. The head of the project told us that previously, they would look at the competitor's product line and see what technology they were utilising and just by seeing the data of the customer they used to launch it with a minor upgradation in the technology; due to this, sometimes their product used to fail when launched in the market. However, this time they employed a design thinking method, and user insights were the primary source of data.

The first step of design thinking is empathy and human-centred approach where user insights are very important. Therefore, employees have themselves gone out to the farmers to examine how they are using rotary tillers and what all problems they are facing. With live interaction with the users, they came to know about so many other problems like changing the rotary rpm as per the tractor or the soil seeds were done manually, and for the sake of changing the rpm, the farmers used to bear 500–700 rupees whenever they must change it. Therefore, the employees came to know as how much it is important to know their customer and their wants and just by sitting in the AC rooms and looking at the data won't give them the valuable insights. So, therefore from this project, one of the most important lessons learnt was to have an interaction with their customer throughout the project and after the prototype of the product was made, to try it with the live customer and taking their feedback and incorporate it.

4.4.2 Project 2 (Case 2)

In this project, the organisation wanted to upgrade their existing product called trailer due to its poor performance in the market. The trailer is a product which is used to carry out spitting operations. After taking customer insights which are the farmers, they came to know that the farmers were only able to use the trailer that is the backward way where they can only empty the material in the backward position but sometimes due to non-availability of space or the position in which they must spit the material changes so therefore they want to spit the material in both sideways positions also. The project team after learning the loophole into their project started working onto the upgradation of the product with ideation tools of design thinking like brainstorming plus the latest technology to produce the prototype which was a three-way spitting trailer where the customer can spit the material in three positions. The prototype was tested on the field with the farmers and with little iteration the product was finally put into the market, and it became a huge success.

4.5 Diffusion Period

After the successful completion of the projects and the product becoming huge success, a report was submitted to the top-level management about the design thinking led project and how it has helped the organisation in upscaling the sales and how they have benefited the employees by learning new capabilities. After getting a nod from the top-level management to further diffuse design thinking into the organisation, many steps were taken. Around 2000 employees were trained where what is design thinking, how it should be used, when to use, in which projects it should be used, when not to use it, when to stop using it, what are the tools and techniques of design thinking, and many more things were taught. It was made mandatory to conduct five training and development session in a year of 1 week duration where employees with their time availability and current role and responsibilities will be selected for the session.

A separate training and development programme was set up for the top-level management and the leaders of the organisation. With this, twenty workshop and conferences were organised where people from different organisation and industry were called to deliver their experience of using design thinking and different tools and techniques were explored. For diffusing design thinking faster in the organisation, a monthly newsletter was launched which contains the applications of design thinking and what recent is going on in design thinking. The success of design led projects was discussed in the newsletter which gives a positive attitude to employees to adopt design thinking in their daily work routine. The design centre which was established only for experimental phase was made permanent, and three more design interns were hired who will mentor the employees and help them in design led projects and handle all the activities of design centre.

In parallel to this, design thinking gradually became an integral part of the organisation culture by introducing it in organisation various ongoing programme like leadership programme, innovation programme, etc. A special programme was designed called design led innovation programme where employees were asked to submit strategic innovation project which uses design thinking as an approach to innovation. The best projects were selected and if the final product of the project became successful in the market, and then the employee was rewarded by sharing 1% profit of that product sales. This gave a boost among employees to use design thinking as their approach. Design thinking approach grew exponentially into the organisation in two ways. First, the projects led by design thinking grew both in number and size, and it was getting embedded into the organisational culture as more and more employees were preferring to use it, and second, user insight was highly valued by the employees which gave boost to design thinking as empathy towards user is the main component of design thinking.

5 Discussion and Implications

There are three ways in which design thinking can be adopted in an organisation, and Table 2 suggests some key recommendations for improving design thinking implementation in the organisation.

5.1 Flexible Implementation

Since design thinking will be implemented in the organisation as per the need of the project in hand, flexible implementation is needed to suit the specific requirement of the project. While adopting this approach, the employees get a hands-on experience as how to use design thinking while working and not just by attending training and development session and after that not knowing how to implement it. It is an approach of learning by doing where you learn and execute the programme. While implementing this approach, the employee capability also gets developed naturally like collaborative, abductive reasoning, etc. The management also gets to know that is this programme is feasible or not in our organisation. If the project yields good result, then further diffusion can be considered. It is like having a practical experience for both employees and management.

5.2 Development of Specialised Unit

Managing design thinking as a separate centre within an organisation allows any new programme to develop new procedures and regulations that are not currently available. It requires manpower, money, and other things to develop a separate centre but starting at a small scale can be beneficial to the organisation. If management truly believes in the concept of any new programme which needs to get implemented, then they can go ahead with this approach. One drawback which needs to be addressed carefully is that because it will be developed as a separate unit, it should be taken into consideration that the centre should engage and aware the whole organisation as

Table 2 Implementation of design thinking through project-based approach

Essential factors	Projects	Key recommendations
Top-level management/ leadership	• Alignment of organisation vision and mission with the outcome of the project • Intent behind using DT conveyed to senior leadership	• Develop strategic vision in collaboration with employees to start a co-create culture • Providing a flexible, free, empowered culture to the employees • Set new key performance indicators keeping in mind the DT culture in the organisation
Availability of resources	• Physical space set up made to provide DT its own space to nurture • 15 Lacs invested in implementing DT for one quarter	• Vital to make long-term investment in design infrastructure, training, and support throughout the company
Human resources initiatives	• Design intern appointed to supervise the DT-backed projects • Developed reward and recognition policies to boost the morale of employees	• Everyone from top leadership to staff members should receive DT training and coaching, in-person workshops, online courses • Making DT as a core competency of the organisation • Emphasise human relationships to foster collaboration and shared goals
Feedback mechanism	• A proper communication level is created with top-level management, human resource department and project team through feedback analysis	• Create a proper channel for employees for sharing their insights

what is happening in the unit. The projects should involve other employees, and the benefits and challenges should be spoken openly through newsletters, workshops, meetings, trainings, etc. openly.

5.3 *Outsourcing*

It is a common practice whenever anything new needs to be implemented in the organisation. With this approach, the organisation will have an outside consultant who will help the organisation with the design led projects. The DT consultants bring the expertise, knowledge, and experience with themselves that the organisation lacks. After the design thinking consultants make the organisation's employees comprehend design thinking, they can leave or be within the organisation as per management decisions. One drawback which design thinking consultants face is that today everyone calls themselves design thinking consultant and to choose the right one can be a problem. The first step towards the implementation of design thinking should be to choose the correct one; otherwise, the organisation will face problem in the long run.

6 Conclusion

This research gives a framework which tells about the essential requirement for the implementation of design thinking and gives three different ways in which it can be achieved. It highlights the important role of human resource in the implementation of design thinking. The firm contact with customers improved because of implementing design thinking. This study looks at how design thinking is introduced, implemented, and integrated within a manufacturing business that is driven by engineering innovation in a sector dominated by product and process advancements. To transition from short-term practices to long-term impact, design thinking necessitates the participation of many individuals. It tells that it is important to be flexible and adapt design thinking to the actual circumstances rather than just adopting the methodology. With the implementation of design thinking, the innovation capabilities of the organisation also increased, but it takes time to build it by adopting the methods, tools, and specially a dynamic mindset.

This is an exploratory study with a limited sample size, so it is more suggestive than definitive. More research is needed to back up the findings. Furthermore, the study sample only talks about the implementation part of design thinking; it can further explore the challenges faced by the manufacturing firm in implementing design thinking. It can further be studied as to how the capabilities developed in employees can be used in their day-to-day activities. Design thinking is a successful way of doing innovation, but it cannot be implemented in each situation. Therefore, this area can be explored as to when to use design thinking and not to use design thinking.

References

1. Bell E, Bryman A (2007) The ethics of management research: an exploratory content analysis. Br J Manag 18(1):63–77
2. Beverland MB, Wilner SJ, Micheli P (2015) Reconciling the tension between consistency and relevance: design thinking as a mechanism for brand ambidexterity. J Acad Mark Sci 43(5):589–609
3. Brown T (2008) Design Thinking. Harv Bus Rev 86(6):85–92
4. Brown T (2009) Change by design: how design thinking transforms organizations and inspires innovation
5. Bucolo S, Wrigley C, Matthews J (2012) Gaps in organizational leadership: linking strategic and operational activities through design-led propositions. Des Manage J 7(1):18–28

6. Dorst K, Cross N (2001) Creativity in the design process: co-evolution of problem–solution. Des Stud 22(5):425–437
7. Holman D, Totterdell P, Axtell C, Stride C, Port R, Svensson R, Zibarras L (2012) Job design and the employee innovation process: the mediating role of learning strategies. J Bus Psychol 27(2):177–191
8. Jaruzelski B, Loehr J, Holman R (2012) Making ideas work. In: PwC Strategy & Global Innovation 1000. Pwc Strategy & Strategy + Business
9. Lafley AG, Martin RL (2013) Playing to win: how strategy really works. Harvard Business Press, Boston
10. Lawson B, Samson D (2001) Developing innovation capability in organisations: a dynamic capabilities approach. Int J Innov Manag 5(03):377–400
11. Lockwood T (2009) Design thinking: integrating innovation, customer experience, and brand value. Allworth Press, New York
12. Martin RL (2011) The innovation catalysts. Harv Bus Rev 89(6):82–87
13. Martin R (2009) The design of business: why design thinking is the next competitive advantage. Harvard Business School Publishing, Boston, MA
14. Osako L, Rocha A, Assi G, Fleury A, Zancul E, Mascia F (2016) Design thinking on the development of an improved concept of toilet usage in airplanes for persons with reduced mobility. In: RESNA/NCART 2016
15. Rauth I, Carlgren L, Elmquist M (2014) Making it happen: legitimizing design thinking in large organizations. Des Manage J 9(1):47–60
16. Roos G (2016) Design-based innovation for manufacturing firm success in high-cost operating environments. She Ji: J Des Econ Inno 2(1):5–28
17. Seidel VP, Fixson SK (2013) Adopting design thinking in novice multidisciplinary teams: the application and limits of design methods and reflexive practices. J Prod Innov Manag 30:19–33
18. Townson P, Matthews J, Wrigley C (2016) Outcomes from applying design-led innovation in an Australian manufacturing firm. Technol Innov Manag Rev 6(6):49–58
19. Walters H (2011) 'Design Thinking' isn't a miracle cure, but here's how it helps. Fast Company (Co. Design blog). Available at http://www.fast-codesign.com/1663480/design-think-insight-a-miracle-cure-but-here-how-it-helps

Effect of Processing Parameters on Surface Roughness, Tool Wear, and Microstructure of Aluminium Alloys

Md Majharul Haque, Kamal Chaudhary, Sayantan Bhattacharya, Vishal Francis, Manjeet Singh, and Narendra Kumar

Abstract Aluminium with one or more additional elements, such as silicon, copper, magnesium, or zinc, makes up the majority of an aluminium alloy. These alloys have excellent corrosion resistance and a high strength-to-weight ratio and outstanding thermal and electrical conductivity, making them widely employed in various industrial and commercial applications. The most promising substance that offers the maximum mechanical strength in the world of hard machining materials is aluminium alloy. The right choice of machining procedure and cooling conditions is crucial when working with aluminium alloy materials. With an emphasis on current advancements and anticipated advances in the industry, we will explore the characteristics, processing, and uses of aluminium alloys in this review paper. The overall objective of this review is to present a thorough overview of the state of the art in aluminium alloy research and to identify key areas for further investigation. We will also discuss the challenges and constraints related to using these alloys.

Keywords Aluminium alloy · Turning · Tool wear · Surface roughness · Microstructure · Built-up edge

1 Introduction

Aluminium alloys are a group of materials that consist primarily of aluminium and one or more other elements, such as Cu, Mg, Si, and Zn. Due to their exceptional capabilities, aluminium alloys are a popular choice for a variety of industrial and commercial applications. The strength-to-weight ratio of aluminium alloys is high, making them both light and durable. They are consequently ideal for use in industries where

M. M. Haque · K. Chaudhary · S. Bhattacharya · V. Francis (✉) · M. Singh
School of Mechanical Engineering, Lovely Professional University, Phagwara, Punjab, India
e-mail: vishal.24813@lpu.co.in

N. Kumar
Department of Industrial and Production Engineering, Dr B R Ambedkar National Institute of Technology, Jalandhar, India

N. Kumar et al. (eds.), *Advances in Materials and Agile Manufacturing*,
Lecture Notes in Mechanical Engineering,
https://doi.org/10.1007/978-981-99-6601-1_7

weight is crucial. They have excellent resistance to corrosion, and aluminium alloys can withstand exposure to harsh environments without degrading or corroding. They are therefore ideal for use both in industrial settings and coastal and marine areas. They have excellent thermal and electrical conductivity, and aluminium alloys can easily dissipate heat and electrical current. Due to their great formability, aluminium alloys may be easily moulded and moulded into a broad variety of shapes and sizes [1]. But working with Al metal may be challenging. Due to their high thermal expansion coefficient, aluminium alloys experience significant thermal expansion and contraction, as a result, dimensional stability. Aluminium alloys are more prone to wear and tear than certain other metals because they have low wear resistance. Due to their poor heat conductivity, significant thermal expansion, and propensity to react with cutting tools, aluminium alloys are challenging to process. When cut or machined, it could even develop a sticky buildup. This occurs as a result of the comparatively low melting temperature of aluminium. Due to the heat of friction, this temperature is so low that it frequently fuses to the cutting edge [2]. Inadequate machining procedures might result in localized overheating in aluminium alloys. The end outcome of alloys machining in post-solution heat treatment tempers is averaging, softening, and subsequently losing strength. In this paper, we have reviewed the properties and challenges of machining aluminium alloys and discussed the various methods, techniques, and tools that are used to enhance these materials' machining capabilities and surface finishes [3]. Since there are several series of aluminium alloys available, each one has a unique capability and set of applications, which are all listed in Table 1.

Table 1 Aluminium series and their applications [4]

Series	Elements alloyed	Ability	Application
1xxx	Pure 99% of Al	High electrical conductivity and corrosion resistance	Electrical conductors, plates, foil, and aircraft parts
2xxx	Cu	High strength and stiffness	Found applications in the aerospace
3xxx	Mg	Good formability and corrosion resistance	Application in beverage cans, attractive foil packaging, and rigid containers
4xxx	Si	Good weldability and formability	Work-hardened rod, sheet, cladding, and forging
5xxx	Mg	High strength and good formability	Rocket cryogenic tanks, rivets, rod, welding, marine, sheet, and vehicle component
6xxx	Mg and Si	Good formability, weldability, and corrosion resistance	Automotive, sheets, marine, forgings, and rod are examples of uses
7xxx	Zn	High strength and good machinability	Aerospace, thick plate, and forging
8xxx	Li	High tensile and yield strength	High temperature aeronautical application, electrical wire

2 Machining of Aluminium Alloy

Due to its weightlessness and high strength qualities, aluminium alloys have seen a substantial increase in application in the industry throughout time. When it comes to better dimensional precision and the greatest surface quality in the sphere of item manufacturing, the machining procedure is frequently the ideal finishing activity. The work surface of the workpiece is extremely exposed during machining practice because of increased temperatures and pressure exposure [5]. Increased metal removal and less burr and built-up edge formation are two benefits of high machining speed.

When machining an aluminium alloy, the feed rate, depth of cut, and cutting speed are the most crucial factors that affect the rate of metal removal [6]. Two cutting tools, chemical vapour-deposited diamond and polycrystalline diamond, have been found to be effective for cutting aluminium alloys in a dry environment [7]. The main difficulty when using an uncoated carbide cemented insert to machine aluminium alloys is the formation of a built-up layer on the rake surface. The work surface's ability to be machined is also influenced by the size and shape of the chips. Aluminium alloys can be machined with high-speed CNC equipment in much less time, which boosts productivity and lowers product costs.

2.1 Turning Operation

Utilizing a cutting tool to remove material from a revolving workpiece to shape or finish an aluminium alloy requires turning. The workpiece is fed into the cutting tool, which is commonly constructed of high-speed steel or carbide, at a precise depth and angle, and it spins while removing the material [8]. Round bars frequently use turning operations for material removal applications. The essential attributes of quality turning machining are good surface roughness profile higher material removal rate, greater dimensional accuracy, and superior tool life. Depth of cut, feed rate, cutting speed, rake angle, size of work, tool geometry, type of material to be processed, vibration of the tool, and type of lubricant used are crucial factors that determine the quality and efficiency of turning operations [9] (Fig. 1).

Due to their distinctive characteristics, many aluminium alloys are difficult to machine. Due to the copper presence, aluminium copper alloys have greater cutting forces and tool wear. The presence of copper can lead to increased tool wear and annealing of the workpiece, which can increase cutting forces and make it more difficult to maintain precise tolerances. Compared to conventional aluminium alloys, aluminium lithium alloys are lighter and have a lower density. Additionally, they are less thermally conductive and have a larger thermal expansion coefficient. This may make it harder to achieve a good surface quality, resulting in higher cutting pressures and greater tool wear. Aluminium silicon alloys frequently gall or seize on cutting

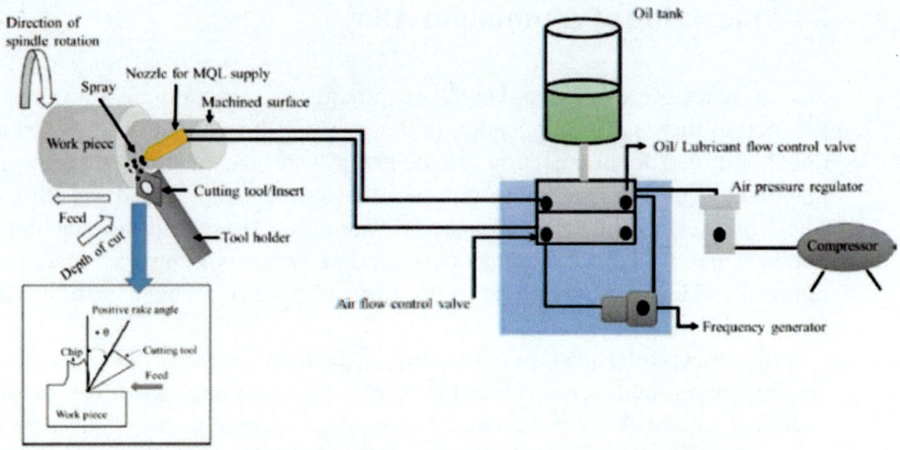

Fig. 1 Schematic of the whole turning operation setup [9]

tools, chipping or breaking them. This could result in more cutting forces and tool wear, and make it difficult to achieve a good surface finish.

To mitigate these issues, we can use a higher cutting speed and feed rate than for pure aluminium alloys and use cutting tools with a positive rake angle, sharp edge, and high-speed steel or carbide material, as well as use a coolant to reduce heat generated during the cutting process [10].

3 Surface Roughness

Due to the impact of high pressures and temperatures during machining, the surface of the machined object is under highly harsh conditions. Al–Li alloys are utilized to enhance the mechanical qualities by improving alloy composition and temper. In comparison to the 2000 and 7000 series alloys, the Al–Li alloys 2090, 2091, 8090, and 8091 have a specific stiffness that is 25% greater and a density that is approximately 10% lower due to the presence of 1.9–2.7% lithium [11]. It is essential to select the best cutting process parameters, such as depth of cut (D), cutting speed (S), and feed rate (F), in order to achieve the desired surface quality. Numerous industrial applications use optimization methods. In order to optimize parameters during machining processes like turning, milling, wire-cut EDM, EDM, and welding, techniques like the Taguchi approach and response surface methodology are used [12]. During machining, the cooling method used has an impact on temperature and BUE formation. As a result, the cooling approach also affects roughness and residual stresses. Applying lubricant can increase tool life and enhance machining performance by lowering process force and built-up edge development [13]. According to some inventions, it is a crucial requirement for machining operations. The surface quality of aluminium alloys after machining may vary depending on the precise alloy composition, the cutting conditions, and the type of cutting tool used; for example, the low ductility of Al–Cu alloys can result in increased tool wear and decreased machining efficiency, the high thermal conductivity of Al–Mg alloys can cause high temperatures at the cutting edge, and the high strength and low ductility of Al–Li alloys can also make them more challenging to machine. To get results in a good surface finish, special cutting tools and coolant systems should be used and also cutting conditions play a very important role. Surface roughness is crucial because it affects the performance of mechanical parts. Surface finish is affected by a variety of elements during turning operations, including the type of workpiece, the feed rate, the cutting tool, the spindle speed, the cut depth, the coolant, the nose radius, the edge angles, and the structure of the tool. It is important to properly explain the connections relationships between the characteristics of the workpiece, the workpiece's properties, and the tool's geometrical parameters. Cutting conditions during machining operations are essential for reducing surface roughness [14]. Tables 2, 3, and 4 shows the surface roughness values using HSS (*H*) and tungsten (*T*) tool with various cutting speed (*S*), feed rate (*F*), and depth of cut (*D*).

For operations involving surface roughness, high-speed CNC machines are used because they are inexpensive, productive, and easy to use. CNC operations include variables that can be controlled, such as feed rate, cutting speed, and depth of cut. To achieve desired surface roughness, the spindle speed, feed rate, and cut depth should be combined in the best possible way using the principal surface method [15].

Table 2 Values of surface roughness at feed rate 0.2 [14]

SN	*S* (m/min)	*F* (mm/rev)	*D* (mm)	*H* (μm)	*T* (μm)
1	300	0.2	1.0	1.20	2.81
2	350	0.2	1.0	0.99	2.57
3	400	0.2	1.0	1.02	2.35
4	450	0.2	1.0	0.93	2.44
5	500	0.2	1.0	0.87	2.23
6	550	0.2	1.0	0.72	2.03
7	600	0.2	1.0	0.61	1.98

Table 3 Values of surface roughness at feed rate 0.4 [14]

SN	*S* (m/min)	*F* (mm/rev)	*D* (mm)	*H* (μm)	*T* (μm)
1	300	0.4	1.0	1.56	3.52
2	350	0.4	1.0	1.43	2.96
3	400	0.4	1.0	1.21	2.62
4	450	0.4	1.0	1.17	2.70
5	500	0.4	1.0	0.90	2.56
6	550	0.4	1.0	0.87	2.39
7	600	0.4	1.0	0.73	2.20

Table 4 Values of surface roughness at feed rate 0.6 [14]

SN	*S* (m/min)	*F* (mm/rev)	*D* (mm)	*H* (μm)	*T* (μm)
1	300	0.6	1.0	1.60	3.61
2	350	0.6	1.0	1.47	3.18
3	400	0.6	1.0	1.28	2.67
4	450	0.6	1.0	1.01	2.60
5	500	0.6	1.0	0.96	2.58
6	550	0.6	1.0	0.89	2.33
7	600	0.6	1.0	0.78	2.19

4 Tool Wear

The machining of aluminium alloy lubricants increases machinability and productivity by decreasing the tool wear and tool life. Dry machining reduces the volume and toxicity of cutting fluids. Cutting fluid reduces friction, removes chips, and decreases the temperature of the tool. Cutting condition with low roughness < 1 μm and tolerance 30 μm for 6 mm of diameter [16]. Tool wear has an effect on the dimensional accuracy, surface quality, tool life, and economics of the machining operation. To ensure the best possible use of cutting tools, various methods for tool wear analysis are a top priority in the machining industries. Using the best possible process parameter combination, predict the cutting tool with 99.7% accuracy for enhanced prediction [17]. Using three different tool materials—uncoated carbide, diamond-coated carbide, and PCD tools—to convert aluminium alloy, a surface polish was created. Compared to uncoated carbide, diamond coating produced a better level of surface polish. Due to its smaller rake angle, the PCD tool produced the maximum radial forces [18], utilizing TP30 and TiN (K10) coated and reinforced aluminium alloy composites. In all cutting situations, increasing cutting speed resulted in a reduction in tool life. When the test was run at a pace of 100 m min1, the tool life for the TP30 tool was 1.96 min, whereas the tool life for the K10 tool was almost 4 min [19] (Figs. 2 and 3).

5 Microstructure

The face-cantered cubic (FCC) phase, sometimes referred to as the alpha phase, is the crystal structure of pure aluminium. The microstructure of the alloy may be changed by minor quantities of additional elements including silicon, copper, magnesium, and zinc, which are present in the majority of alloys made of aluminium. Al alloys often have the following microstructures. Low alloying content alloys often have a single-phase microstructure, which is distinguished by a uniform distribution of the main phase. Two-phase microstructure: This is defined by the existence of a primary phase

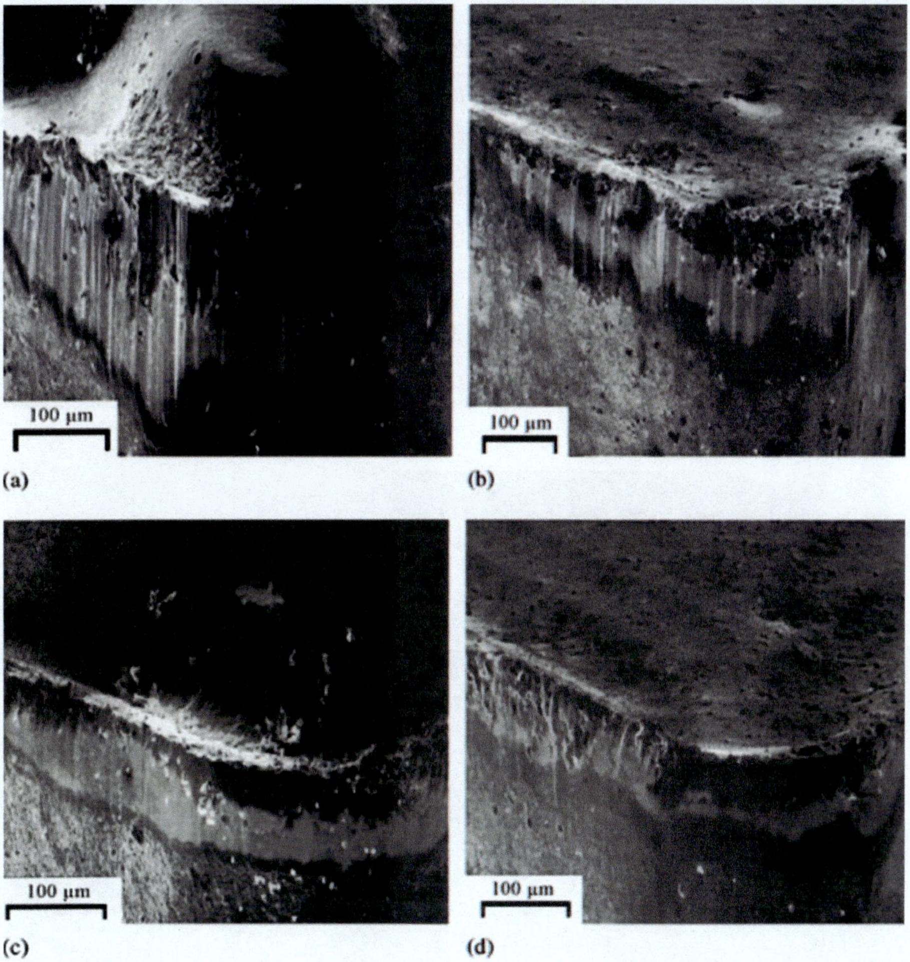

Fig. 2 Flank wear of coated carbide cutting tools when machining of the 30 wt% Al_2O_3 particle-reinforced composite [19]

and a second phase and is present in alloys with modest levels of alloying. Multi-phase microstructure: This is defined by the presence of many phase boundaries and is present in alloys with substantial alloying content. An Al alloy's microstructure may be impacted by machining, particularly if the alloy has been heat-treated beforehand. The heat produced by the cutting tools during machining may raise the workpiece's temperature. Recrystallization and grain growth may result from this, altering the alloy's microstructure. Additionally, machining-related plastic deformation can alter the microstructure of a material. To minimize the effect of machining on the microstructure of Al alloys [20], investigations have been made into the cut surfaces' microstructures and ranges of hardness produced by various processes. The type of cutting process has an impact on the cut surface's hardness and surface finish.

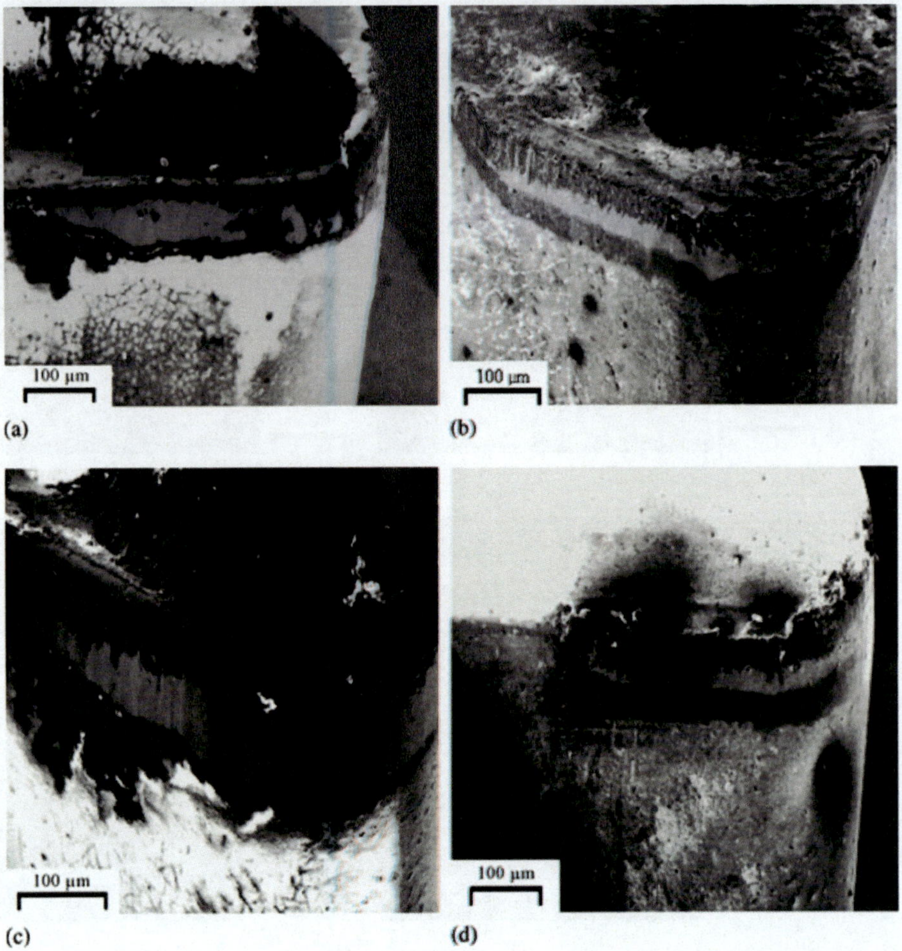

Fig. 3 Flank wear of coated carbide cutting tools when machining of the 20 wt% Al_2O_3 particle-reinforced composites [19]

The type of cutting process has an impact on the microstructure of cut surfaces. All cutting processes, excluding abrasive water jet, cause microstructural changes in the materials being cut. The abrasive water jet technique can be successfully used in industrial applications where no microstructural changes and hardness reduction are required [21] (Fig. 4).

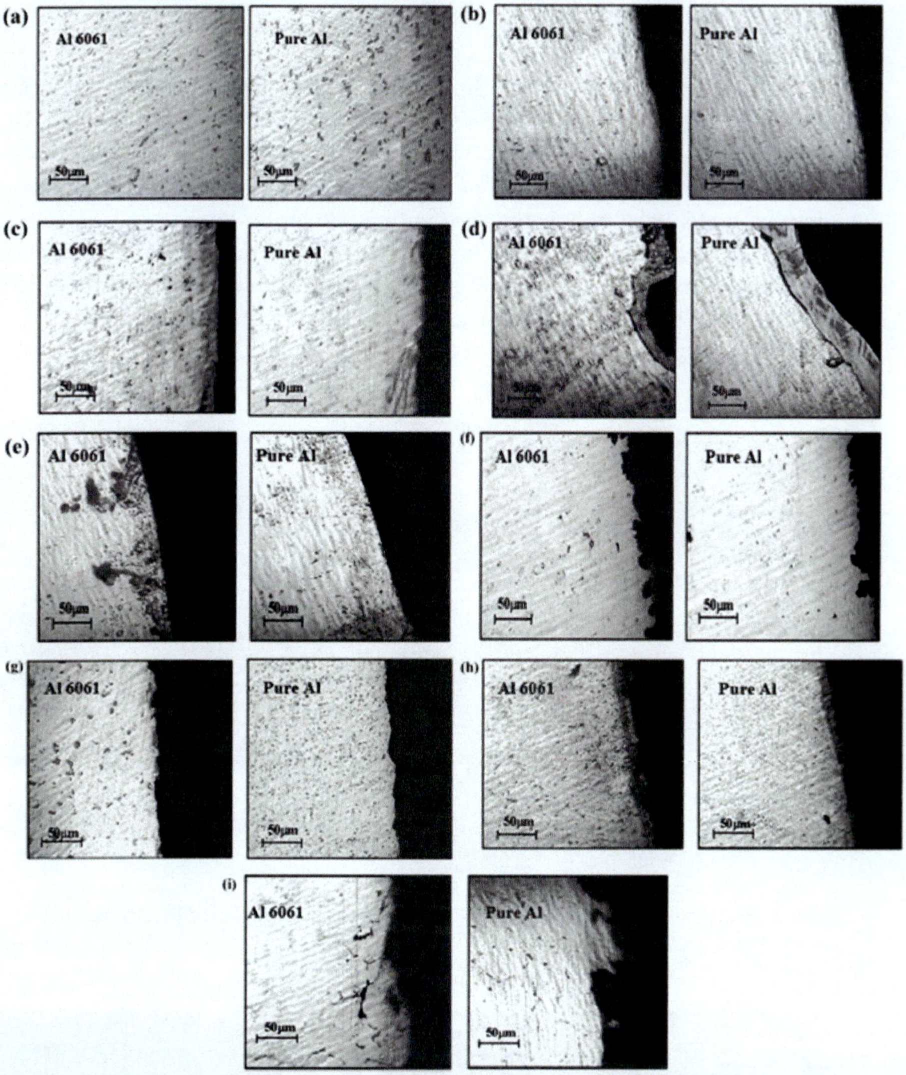

Fig. 4 Effect of the machining process on the microstructure of aluminium and aluminium alloy [21]

6 Built-Up Edge

The quality of the workpiece is impacted by BUE formation and evolution. Mechanical adhesion causes BUE to form. A significant amount of stress and temperature builds up in the secondary deformation zone at the chip tool interface when machining ductile metals like steel with long chip tool contact lengths. BUE are typically caused by a variety of mechanisms that primarily depend on the materials of the tool and the workpiece [22] (Figs. 5 and 6).

BUE formation may affect the quality of the surface because the affixed work materials' can pass beneath the tool flank face and interact with it. The BUE edge also changes the tool rake angle, which has an impact on the morphology of the chips. At the tool-chip interface, e BUE can frequently form under high friction conditions, and the stress level in the area surrounding the tool's cutting edge has a big impact on e BUE's morphology. Orthogonal cutting tests were performed in dry conditions on a planer machine with a slow cutting speed to observe the cutting process (chip formation) with a CCD camera, and on a computer numeric control (CNC) lathe to allow variation in the cutting speed in a broad range [23] (Fig. 7).

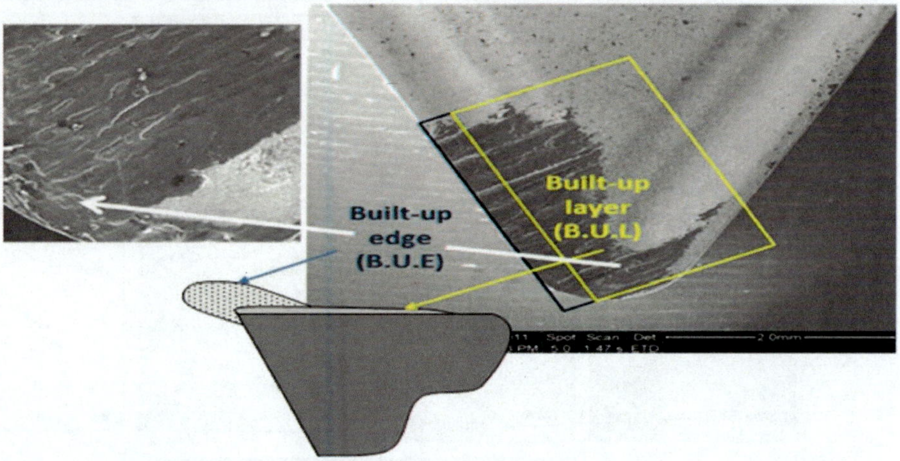

Fig. 5 SEM image of a turning insert's edge and tool rake face after dry machining the UNS A97050 alloy [22]

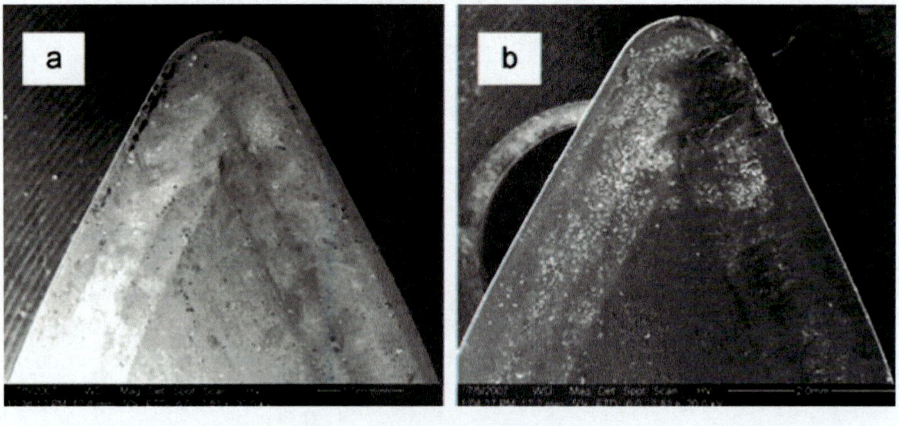

Fig. 6 SEM image of **a** uncoated and **b** TiN coated tools after dry turning UNS A97050 [22]

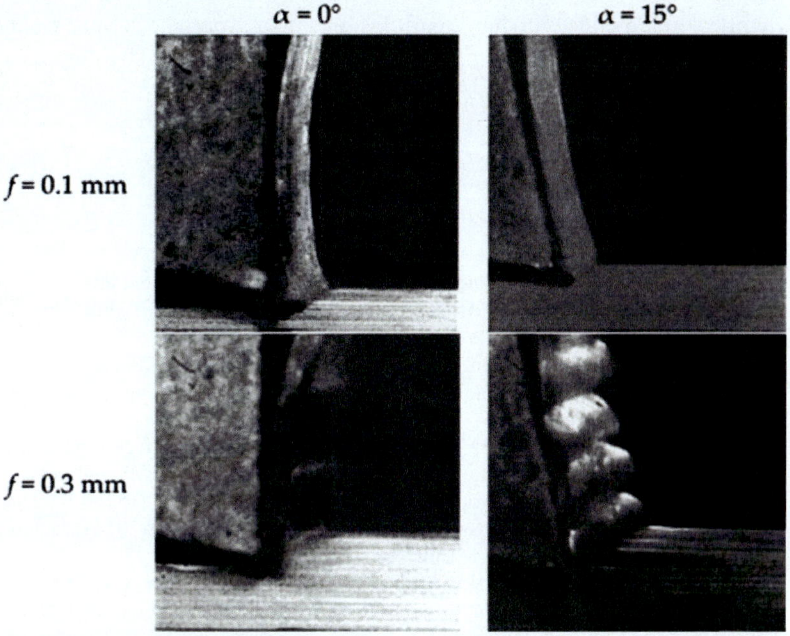

Fig. 7 Chip morphology measured at a 60 m/min cutting speed [23]

The BUE happens while machining ductile metals like the aluminium alloy AA2024-T351 under precise cutting circumstances. On an instrumented planar machine, the impact of uncut chip thickness on the generation of the BUE was investigated while the tool rake angle was changed. The CCD camera recorded a variety of cutting sequences, which were then examined. Instantaneous photographs in Fig. 7 demonstrate that the BUE occurs when the chip thickness is 0.1 mm and the rake angle is low (i.e. 0°) [23].

7 Conclusion

Machining of aluminium alloys can present challenges due to their low thermal conductivity, high reactivity with cutting tools, and tendency to work-harden. However, with the proper selection of cutting tools and machining parameters, aluminium alloys can be machined efficiently and effectively. The machinability of aluminium alloys can be increased through the use of coated cutting tools and high-speed machining procedures. Additionally, proper tool path planning and chip control can also help prevent issues such as built-up edge and heat-affected zones. High-speed steel, N-grade cemented carbide, and synthetic diamond are the most suggested materials for tools. Depending on the alloy and the process, the proposed geometries have a very high probability of success. The kind of wear is flank wear, while some metals with hard particles may also experience crater wear or even notch wear. All of the properties of the material will be favoured by the use of cutting fluid in the form of flood cooling or MQL.

References

1. Handbook ASM (1989) Machining of aluminium and aluminium alloy
2. Santos Jr MC, Machado AR, Sales WF (2016) Machining of aluminium alloys: a review. Springer-Verlag London
3. List G, Nouari M, Géhin D, Gomez S, Manaud J-P, Le Petitcorps Y, Girot F (2005) Wear behaviour of cemented carbide tools in dry machining of aluminium alloy. Wear 259(7–12):1177–1189
4. Samuel AU, Araoyinbo AO, Elewa RR, Biodun MB (2021) Effect of machining of aluminium alloys with emphasis on aluminium 6061 alloy–a review. In: IOP conference series: materials science and engineering, vol 1107, no 1. IOP Publishing, p 012157
5. Santos MC, Machado AR, Sales WF, Barrozo MA, Ezugwu EO (2016) Machining of aluminum alloys: a review. Int J Adv Manuf Technol 86:3067–3080
6. Deepak D, Rajendra B (2015) Studies on material removal rate of Al 6061 while turning with coolant and without coolant using Taguchi method. Int J Res Eng Technol 4(09):75–78
7. Soren TR, Kumar R, Panigrahi I, Sahoo AK, Panda A, Das RK (2019) Machinability behavior of aluminium alloys: a brief study. Mater Today: Proc 18:5069–5075
8. Zakaria MS, Afiq A, Hafiz M, Jamalludin MR, Rosli MU, Rahim WM, Ishak MI, Khor CY, Nawi MAM, Shahrin S (2018) Turning of aluminum alloy Al-6063 under dry and wet condition using CVD coated tool. In: AIP conference proceedings, vol 2030, no 1. AIP Publishing
9. Muhammad A, Sayd L, Akhtar R, Khalid QS, Khan AM, Pruncu CI (2020) Optimization of machining parameters of aluminum alloy 6026-T9 under MQL-assisted turning process. J Mater Res Technol 9(5):10916–10940
10. Batista M, Del Sol I, Gomez-Parra A, Ramirez-Peña M, Salguero J (2019) Study of the tool wear process in the dry turning of Al–Cu alloy. Metals 9(10):1094
11. Dursun T, Soutis C (2014) Recent developments in advanced aircraft aluminium alloys. Mater Des (1980–2015) 56:862–871
12. Vinoth V (2021) Optimization of process parameters in turning of aluminum alloy using response surface methodology. Mater Today: Proc 46:9462–9468
13. Junge T, Mehner T, Nestler A, Schubert A, Lampke T (2022) Surface properties in turning of aluminum alloys applying different cooling strategies. Procedia CIRP 108:246–251
14. Lawal SA, Ahmed AM, Lawal SS, Ugheoke BI (2016) Effect of HSS and tungsten carbide tools on surface roughness of aluminium alloy during turning operation
15. Abdallah A, Rajamony B, Embark A (2014) Optimization of cutting parameters for surface roughness in CNC turning machining with aluminum alloy 6061 material. Optimization 4(10):1–10
16. Nouari M, List G, Girot F, Coupard D (2003) Experimental analysis and optimisation of tool wear in dry machining of aluminium alloys. Wear 255(7–12):1359–1368
17. Okokpujie IP, Ikumapayi OM, Okonkwo UC, Salawu EY, Afolalu SA, Dirisu JO, Nwoke ON, Ajayi OO (2017) Experimental and mathematical modeling for prediction of tool wear on the machining of aluminium 6061 alloy by high-speed steel tools. Open Eng 7(1):461–469
18. Reis DD, Abrao AM (2005) The machining of aluminium alloy 6351. Proc Inst Mech Eng, Part B: J Eng Manuf 219(1):27–33
19. Sahin Y, Kok M, Celik H (2002) Tool wear and surface roughness of Al_2O_3 particle-reinforced aluminium alloy composites. J Mater Process Technol 128(1–3):280–291

20. Oladapo BI, Abolfazl Zahedi S, Omigbodun FT, Oshin EA, Adebiyi VA, Malachi OB (2019) Microstructural evaluation of aluminium alloy A365 T6 in machining operation. J Mater Res Technol 8(3):3213–3222
21. Akkurt A (2015) The effect of cutting process on surface microstructure and hardness of pure and Al 6061 aluminium alloy. Eng Sci Technol Int J 18(3):303–308
22. Gómez-Parra A, Álvarez-Alcón M, Salguero J, Batista M, Marcos M (2013) Analysis of the evolution of the built-up edge and built-up layer formation mechanisms in the dry turning of aeronautical aluminium alloys. Wear 302(1–2):1209–1218
23. Haddag B, Atlati S, Nouari M, Moufki A (2016) Dry machining aeronautical aluminum alloy AA2024-T351: analysis of cutting forces, chip segmentation and built-up edge formation. Metals 6(9):197

Fabrication of Functionally Graded Materials: A Review

Ashutosh Kumar Gupta and Mohammad Taufik

Abstract Functionally graded materials (FGMs) are a class of advanced materials engineered with a unique property gradient. In the functionally graded materials (FGMs) constituent and microstructures slowly changes along spatial directions in the various condition, so that change in functions and properties. FGMs can be personalized for enhanced performance. Functionally graded materials can be manufactured using various processing methods but additive manufacturing provides higher degree of control over spatial resolution and precise pathway than presently existing techniques. Since, additive manufacturing (AM) deposits materials in layer-by-layer manners of single or multiple materials, so AM has potential to regulate the composition and microstructure in multiple dimensions and directional gradient structures. This paper addresses the modeling, processing, and microstructures of FGMs produced by conventional and AM processes.

Keywords Functionally graded materials · Composite · Direct energy deposition · Additive manufacturing · Electron beam melting

1 Introduction

Functionally gradient materials (FGMs) are gaining attention in the field of modern material processing because incremental changes in the composition and structure over volume result, gradual changes in the material's properties to satisfy the requirements [1]. FGMs can be tailored to particular functions and applications as they can incorporate two or more materials into a single framework to perform various tasks depending upon the environment condition [2]. The development of modern day FGMs was first proposed in the 1980s in Japan for the use in rocket engine because traditional composite materials employed in this project failed repeatedly

A. K. Gupta · M. Taufik (✉)
Department of Mechanical Engineering, Maulana Azad National Institute of Technology, Bhopal, India
e-mail: mohammad.taufik@manit.ac.in

N. Kumar et al. (eds.), *Advances in Materials and Agile Manufacturing*,
Lecture Notes in Mechanical Engineering,
https://doi.org/10.1007/978-981-99-6601-1_8

due to delamination of the metal and ceramic components used. The failure of the composite used was blamed on the presence of a sharply defined interface between the materials. The reason behind this sharply defined interface is because the two components perceive each other as distinct materials, making it easy for them to deboned while operating under harsh and extreme loading condition. Imagine a situation in which one material is progressively added to another, so that the one, or base, substance only views the second as an impurity spread inside it. Figure 1a shows the schematic of various type of functionally graded material, and Fig. 1b shows example of FGMs present in nature (the human bone) [3]. The FGMs have lower residual stresses so the improved mechanical and physical properties are offered by FGMs parts [4, 5]. FGMs can be classified into four categories based on distribution of material gradients (1) fraction gradient, (2) shape gradient, (3) orientation gradient, (4) size gradient which are also shown in Fig. 1a, respectively.

The conventional manufacturing processing techniques include powder metallurgy, plasma spraying, chemical vapor deposition, and Galvano forming. These older FGMs processing techniques have been re-evaluated and changed, and new ones have been developed, from 1991 to the present. Recent advancements in additive manufacturing (AM) techniques allow the printing of components and structures.

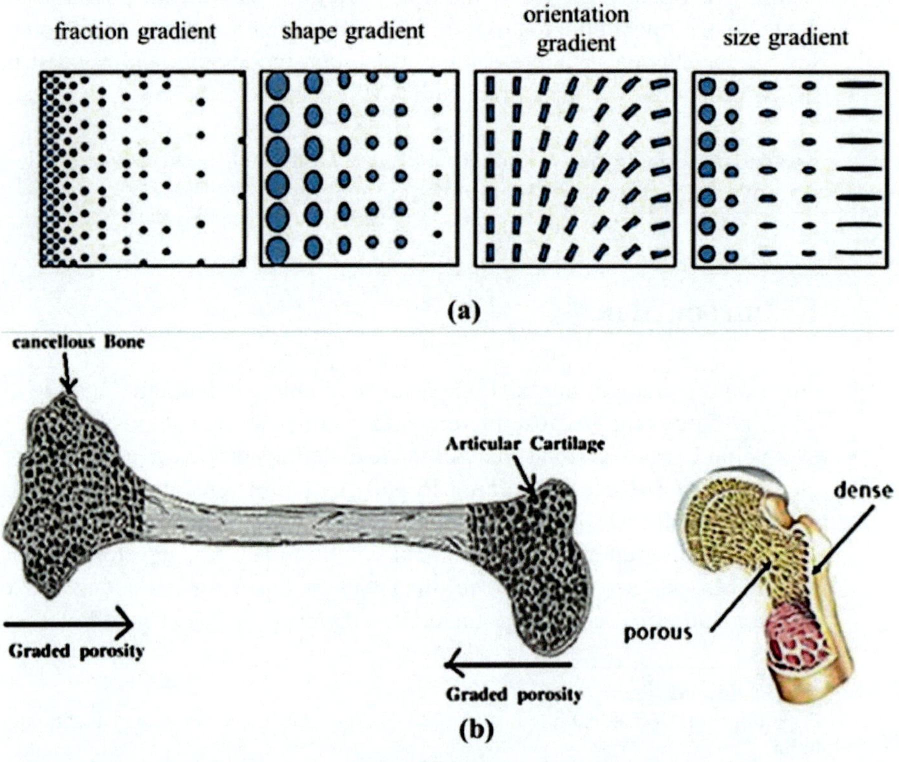

Fig. 1 **a** Various types of FGMs, **b** natural presence of FGMs bone [3]

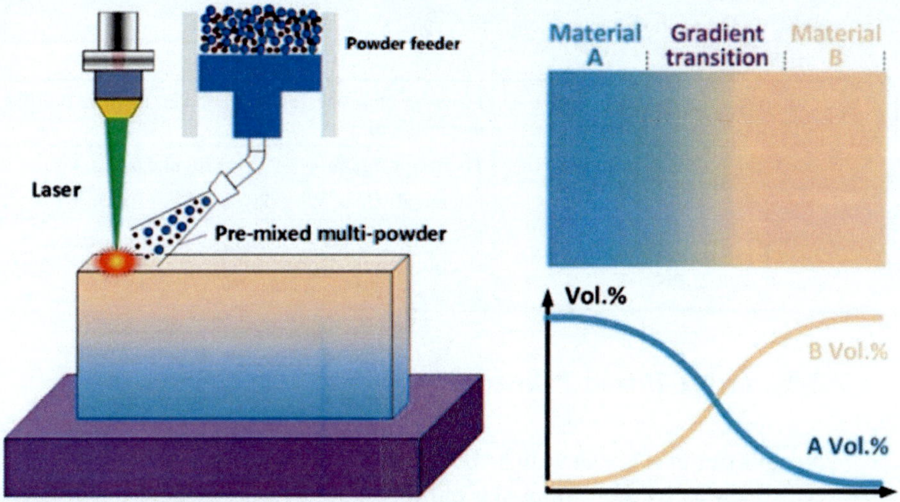

Fig. 2 Schematic diagram to produce the FGM via DEP [9]

Metallic components and structures can be printed by powder processing techniques, i.e., selective laser melting and selective electron beam melting processes without any special tooling using a CAD file [6–8]. Wire-based AM processes are limited to polymers and not able to print net shape filigree structures, so the researchers are concentrating on powder processing techniques.

Figure 2 is the schematic diagram of directed energy deposition to fabricate FGMs. LMT may be used to create complicated FGM components that suit the best technical design for a variety of applications; the performance of components produced by AM methods is similar as high-performance die-casting equipment [9]. A number of alloys like nickel-based superalloys [10], titanium alloys [1, 11], aluminum alloys [12], and steels [13] have been processed by additive manufacturing methods in previous years. Since FGMs show variable properties that enable them to endure severe working environments, they can be used in military and aerospace [11].

2 Fabrication of Functionally Graded Materials (FGMs)

In the recent two decades, there has been a significant advancement in the development of FGMs production methods, resulting in many FGMs manufacturing methods, as shown in Table 1. The type of raw materials, functional qualities, and facility availability required lead the implementation of these production procedure, which comes in mechanical and chemical forms.

Table 1 Fabrication methods and their types

Method	Type(s)
Additive manufacturing (AM)	Laser-based process, electron beam melting process, material jetting
Melting process	Thermal spray process, centrifugal casting
Vapor deposition (VD) process	Chemical VD (CVD), physical VD (PVD)
Powder metallurgy	Power centrifugal forming, vacuum slip casting, electrophoretic deposition, wet powder spraying, stepwise compositional control

2.1 Laser-Based Process

Niendorf et al. [7] stated that the microstructure evolution is dominated in a very effective way by the various sets of process variables. The experiments were carried out on AISI 316L material (particle size = 40 μm) using selective laser melting technique at two laser power 400 and 1000 W. The range of layer thickness obtained using the SLM process was 50–150 μm because layer thickness depends upon the laser intensity. In order to obtain the final shape of the structure, first, the fabricated part was heat treated for two hours at 650 °C under argon atmosphere and after that part was cut by using an EDM process. Texture and grain size are the reason for different μ-structure present in 316L steel Fig. 3. The analyses reveal that columnar coarse-grained material with strong texture shows yield strength, ultimate strength, and Young's modulus low values. On the other hand, fine-grained material with strong texture shows higher strength and Young's modulus values.

Mahamood and Akinlabi [11] used laser metal deposition technique to develop functionally graded Ti6Al4V/TiC composite. For the manufacturing of FGMs composite, a 5-mm-thick sheet made of Ti6Al4V was set as base plate and powder of Ti6Al4V and TiC were used. Later, the properties of the base plate (i.e., Ti6Al4V plate) were compared with the FGMs composite which was fabricated at fixed parameters. The FGMs composite shows four times better-improved value of microhardness than microhardness of base material.

Beal et al. [13] investigated the performances of pard made of FGMs using selective laser fusion process. Three combinations of pre-mixed H13 and Cu powder with various laser scan strategies were analyzed. In order to surge powder packing of spherical particles, bimodal distributions were made particle size ratio to around 7:1. Four different laser scanning methods; filling, sequential, alternated, and refill were used to quickly complete the scan. The pure H13 powder offered a regular structure, but it is conceived those internal stresses were preserved in the parts because it solidified firstly from supercritical temperatures. It was found that the refill strategy is the most suitable to produce denser parts.

Zhang et al. [14] successfully manufactured Ti6Al4V/TiC FGM by direct laser deposition (DLD) process using Ti6A4V and TiC powder and investigated the effect of laser power, scan speed, and TiC percentage composition on mechanical property,

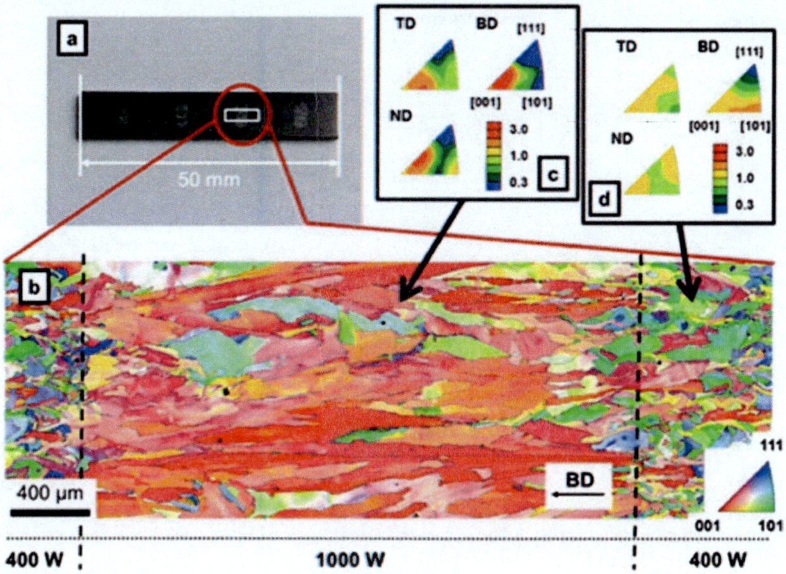

Fig. 3 Microstructural analysis of a sample portion [7]

hardness, and microstructure. Relationships between the process parameters (i.e., laser power, scan speed, and %composition of TiC) with the responses (i.e., hardness, stress–strain, and UTS) are combined in Fig. 4 SEM analysis shows that size and shape of TiC particles in FGMs were different from original TiC powder particles, because of dissolution and solidification of TiC particles.

Lee et al. [9] concluded that pre-mixed multiple powders directly affect the composition of functionally graded materials produced by direct energy deposition method Fig. 5a Pure copper powder and aluminum 4047 powder were used to perform two experiments at different particle sizes. For experiment 1, ratio of dia. of Al particles to dia. of copper particles (DAl-P/DCu-P) is 1.12, and for experiments 2, dia. of Al particles to dia. of copper particles (DAl_P/DCu_P) is 1.854 which was taken. The experimental design idea and two solidified epoxy resin layers containing PMM-powder patterns from both experiments are shown in Fig. 5b. For continuous flow, an ideal value was fixed as square root of powder density ratio.

2.2 Electron Beam Melting Process

The electron beam melting method for fabrication of FGMs is very complex, but the manufacturing procedure can be divided into four sections. Firstly, apply one layer of appropriate mixture ratio of powder on the preheated base plate. Secondly, preheat the powder bed to prevent powder scattering (i.e., formation of bond the powder particles). Thirdly, use the focused electron beam to melt the powder bed and to 2D

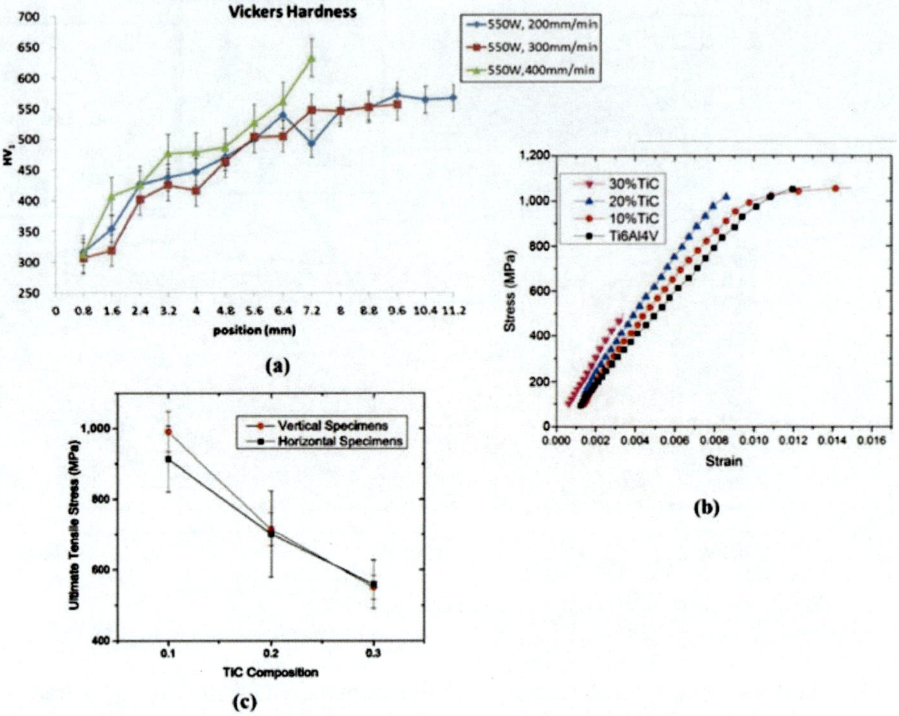

Fig. 4 **a** Variation of hardness at various laser power and scan speed. **b** Relationship curves of stress–strain at various % of TiC. **c** Variation of UTS with TiC compositions [14]

cross-section of required 3D parts. Lastly, move the base plate by one layer thickness (Fig. 6) [2].

Zhou et al. [15] used electron beam selective melting (EBSM) process to print the 1D FGMs of TiAl alloy ($\alpha_2 + \gamma$) and Ti alloy ($\alpha + \beta$). For the manufacturing of 1D FGMs parts, Ti47Al2Cr2Nb alloy (particle size: 50–108 μm), electron gun (accelerating voltage 60 kV), and small forming cylinder (size: 100 × 100 × 100 mm) at the TC4 alloy sheet (ethanol cleaned) setup used. ASTM E8 was used to produce the tensile specimens, and the testing was performed at the room temperature at 1.2×10^{-3}/s strain rate on Instron tensile testing machine. After that, author used SEM testing to examine the fracture behavior of test specimens. Also, the effect of the overlap scan distance was evaluated on the microstructure, microhardness, and tensile properties.

Hrabe and Quinn [16] investigated the two parameters, namely distance from build plate and part size, and the tensile testing, microhardness, and the microstructure evaluation were performed for the each fabricated specimen. The authors found that distance from the build plate was not of any response; it is because of the preheating of build plate. However, part size has significant effect on the responses but not more than 2% over the size range. Fabrication of FGMs is also reported in Juechter et al. [17] Ge et al. [18], H, etc.

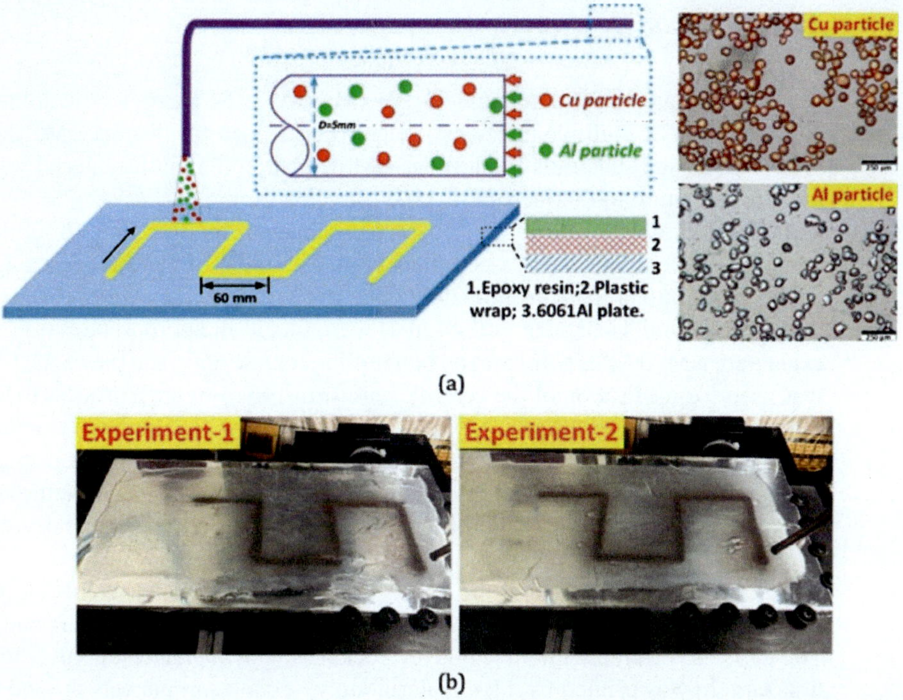

Fig. 5 **a** Schematic of experimental setup, **b** produced parts [9]

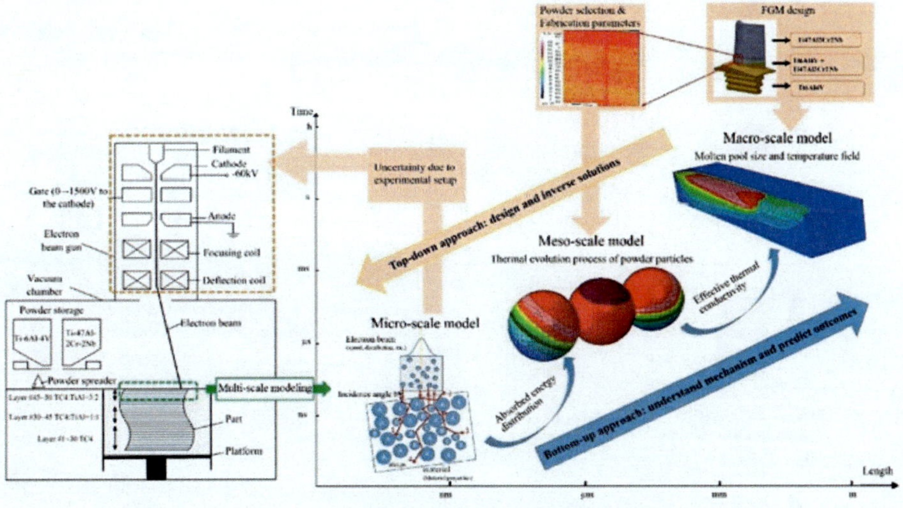

Fig. 6 Schematic diagram of proposed multi-scale modeling framework [2]

2.3 Material Jetting

Material jetting additive manufacturing (MJAM) uses piezoelectric inkjet heads which consist of multiple nozzles to selectively deposit the droplets. MJAM is also known as digital materials because MJAM deposited the multi-material in discrete and digital nature to fabricate parts [19, 20].

Levy et al. [21] used binder jet AM technique for the first time to fabricate the complex-shaped part of TiCx ceramics with a gradient of carbon percentage (0.7–0.98). In order to prepare the non-stoichiometric TiCx powders, first TiC0.98 (P_size = 0.44 μm) and Ti (P_size = 0.44 μm) were mixed in appropriate ratio (equation evaluated in MATLAB software) by ball milling for one hour, and then SAE1070 steel was used for infiltration of the ceramics preforms. Step-by-step process to fabricate a part using binder jet AM technique is shown in Fig. 7. There is no evidence of free TiC in heat-treated parts, which was confirmed by X-ray diffraction analysis. The analysis results show that μ-structure is mostly consistent with untransformed alpha-ferrite on the TiC0.7 side. Fine martensite μ-structure was clearly observed on TiC0.98 side.

Salcedo et al. [20] validated the four FGMs parts made with different compositions by MJAM technique and compared them using finite element analysis and testing. The parts are printed with 30 μm layer thickness with standard support setting. The first sample was printed by TB+ material only, second sample was printed by TB+

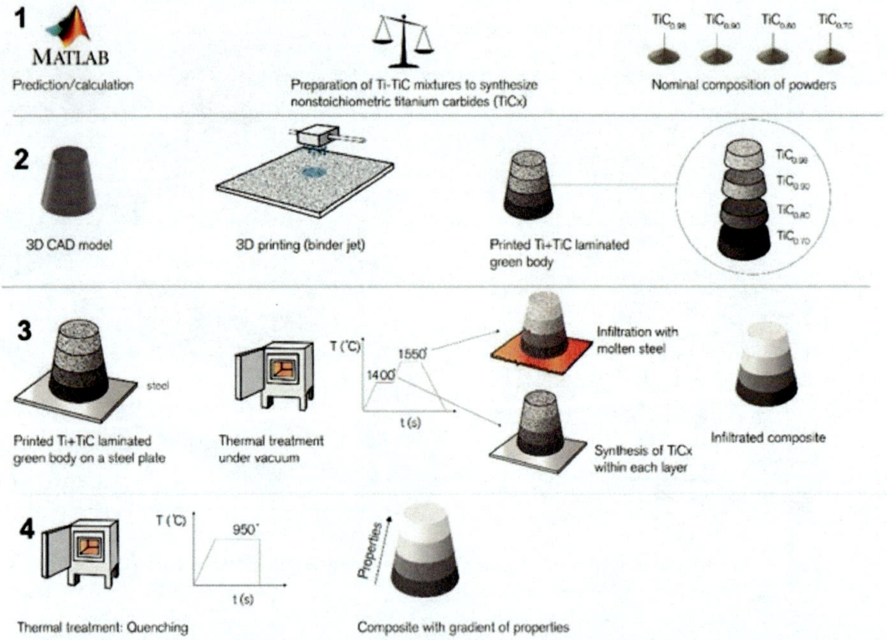

Fig. 7 Manufacturing of 3D printed part using graded TiC/steel FGMs [21]

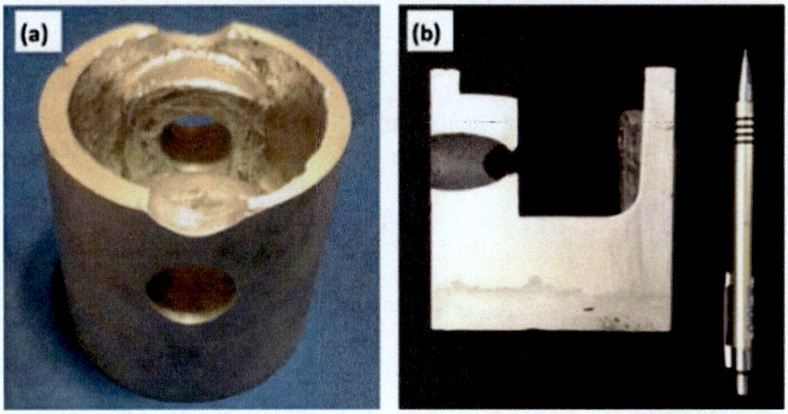

Fig. 8 **a** Centrifugally casted piston Si reinforced A390 + 0.5%Mg FGM composite, **b** section; vertical cut cross-sectional of A390 FGM piston [22]

with an addition of VW+, third sample was printed by using the combination of TB+ and DM95, VW+ with circular graded islands scan, and fourth sample was printed by using the combination of TB+ and DM95, VW+ with rectangular graded islands scan. The results show that FEM analysis can precisely predict the testing result, and elongation of FGMs sample is enhanced by 170% in the TB+ area.

2.4 *Conventional Techniques*

Arsha et al. designed, fabricated, and evaluated the functionally graded automotive piston made of in situ primary silicon (Si) reinforced A390 and A390-0.5% Mg composite produced by using centrifugal casting technique (Fig. 8). Microstructure, mechanical property, thermal property, and chemical property are characterized along the vertical section of fabricated piston head and compared with part produced gravity die casting. The comparison result shows that FGMs piston has graded distribution of Si from head region to skirt portion, higher hardness in head section, outer periphery has lower wear rate, and head can survive in high temperature with little distortion.

3 Discussion, Conclusion, and Future Scope

The main objective of this paper was to review and introduce the field of FGMs and the fabrication of FGMs using laser melting, electron beam melting, powder fusion, and conventional techniques. Laser-based process and electron beam melting techniques are the AM techniques which are gaining popularity in the fabrication of FGMs among the researchers. The literature survey shows that AM methods may be used to fabricate the complex part with almost any possible combination of powders alloys. A lot of process variables like power intensity, powder particle size, powder feed rate, and powder percentage combination affect the characteristics of produced parts. In closing, dimensional accuracy, surface roughness, and process optimization generally have not been carried out for the produced part of FGMs techniques.

Acknowledgement This work was supported by the Science and Engineering Research Board (SERB)—DST under its Start-up Research Grant (SRG) scheme [Grant number: SRG/2019/000943].

References

1. Bobbio LD, Bocklund B, Otis R, Borgonia JP, Dillon RP, Shapiro AA et al (2018) Characterization of a functionally graded material of Ti-6Al-4V to 304L stainless steel with an intermediate V section. J Alloys Compd 742:1031–1036. https://doi.org/10.1016/j.jallcom.2018.01.156
2. Yan W, Ge W, Smith J, Lin S, Kafka OL, Lin F et al (2016) Multi-scale modeling of electron beam melting of functionally graded materials. Acta Mater 115:403–412. https://doi.org/10.1016/j.actamat.2016.06.022
3. Ebhota WS (2018) Casting and applications of functionally graded metal matrix composites. In: Vijayaram T-CJE-TR (ed) IntechOpen, Rijeka, p Ch 5. https://doi.org/10.5772/intechopen.71225
4. Jackson TR, Liu H, Patrikalakis NM, Sachs EM, Cima MJ (1999) Modeling and designing functionally graded material components for fabrication with local composition control. Mater Des 20:63–75. https://doi.org/10.1016/S0261-3069(99)00011-4
5. Stansbury JW, Idacavage MJ (2016) 3D printing with polymers: challenges among expanding options and opportunities. Dent Mater 32:54–64. https://doi.org/10.1016/j.dental.2015.09.018
6. Taufik M, Jain PK (2017) Laser assisted finishing process for improved surface finish of fused deposition modelled parts. J Manuf Process 30:161–177. https://doi.org/10.1016/j.jmapro.2017.09.020
7. Niendorf T, Leuders S, Riemer A, Brenne F, Tröster T, Richard HA et al (2014) Functionally graded alloys obtained by additive manufacturing. Adv Eng Mater 16:857–861. https://doi.org/10.1002/adem.201300579
8. Gupta AK, Taufik M (2021) Effect of process variables on performances measured in filament and pellet based extrusion process. Mater Today Proc 47:5177–5184. https://doi.org/10.1016/j.matpr.2021.05.508
9. Lee J-Y, An J, Chua CK (2017) Fundamentals and applications of 3D printing for novel materials. Appl Mater Today 7:120–133. https://doi.org/10.1016/j.apmt.2017.02.004
10. Pollock TM, Tin S (2006) Nickel-based superalloys for advanced turbine engines: chemistry, microstructure and properties. J Propuls Power. https://doi.org/10.2514/1.18239
11. Mahamood RM, Akinlabi ET (2015) Laser metal deposition of functionally graded Ti6Al4V/TiC. Mater Des 84:402–410. https://doi.org/10.1016/j.matdes.2015.06.135
12. Li W, Zhang J, Zhang X, Liou F (2017) Effect of optimizing particle size on directed energy deposition of functionally graded material with blown pre-mixed multi-powder. Manuf Lett 13:39–43. https://doi.org/10.1016/j.mfglet.2017.07.001
13. Beal VE, Erasenthiran P, Hopkinson N, Dickens P, Ahrens CH (2006) The effect of scanning strategy on laser fusion of functionally graded H13/Cu materials. Int J Adv Manuf Technol 30:844–852. https://doi.org/10.1007/s00170-005-0130-x
14. Zhang J, Zhang Y, Li W, Karnati S, Liou F, Newkirk JW (2018) Microstructure and properties of functionally graded materials Ti6Al4V/TiC fabricated by direct laser deposition. Rapid Prototyp J 24:677–687. https://doi.org/10.1108/RPJ-12-2016-0215

15. Zhou J, Li H, Yu Y, Firouzian K, Qian Y, Lin F (2020) Characterization of interfacial transition zone of functionally graded materials with graded composition from a single material in electron beam powder bed fusion. J Alloys Compd 832:154774. https://doi.org/10.1016/j.jallcom.2020.154774
16. Hrabe N, Quinn T (2013) Effects of processing on microstructure and mechanical properties of a titanium alloy (Ti–6Al–4V) fabricated using electron beam melting (EBM), part 1: distance from build plate and part size. Mater Sci Eng A 573:264–270. https://doi.org/10.1016/j.msea.2013.02.064
17. Juechter V, Scharowsky T, Singer RF, Körner C (2014) Processing window and evaporation phenomena for Ti–6Al–4V produced by selective electron beam melting. Acta Mater 76:252–258. https://doi.org/10.1016/j.actamat.2014.05.037
18. Ge W, Lin F, Guo C (2015) Functional gradient material of Ti-6Al-4V and γ-TiAl fabricated by electron beam selective melting. In: International solid freeform fabrication symposium, vol 602. Tsinghua University, p 613
19. Ituarte IF, Boddeti N, Hassani V, Dunn ML, Rosen DW (2019) Design and additive manufacture of functionally graded structures based on digital materials. Addit Manuf 30:100839. https://doi.org/10.1016/j.addma.2019.100839
20. Salcedo E, Baek D, Berndt A, Ryu JE (2018) Simulation and validation of three dimension functionally graded materials by material jetting. Addit Manuf 22:351–359. https://doi.org/10.1016/j.addma.2018.05.027
21. Levy A, Miriyev A, Elliott A, Babu SS, Frage N (2017) Additive manufacturing of complex-shaped graded TiC/steel composites. Mater Des 118:198–203. https://doi.org/10.1016/j.matdes.2017.01.024
22. Arsha AG, Jayakumar E, Rajan TPD, Antony V, Pai BC (2015) Design and fabrication of functionally graded in-situ aluminium composites for automotive pistons. Mater Des 88:1201–1209. https://doi.org/10.1016/j.matdes.2015.09.099

A Review on the Fused Filament Fabrication of Lightweight Components

Sonika Sahu and Pradeep Kumar

Abstract Fused filament fabrication (FFF) has become a de facto choice to fabricate complex-shaped polymer components. Recently, the potential of this technique has been explored in the fabrication of lightweight components due to its capabilities. Various articles have been reported based on the use of FFF process for the development of lightweight structures. In order to identify the current status, challenges, and research gaps in this domain, a thorough review of the existing articles is needed. Therefore, the present article aims to review the development of lightweight components through FFF. Different strategies adopted to develop lightweight components are highlighted with the challenges and research gaps. The outcome of the study may be useful in developing components for weight-sensitive applications using FFF process.

Keywords Additive manufacturing · FFF · Lightweight components · Weight-sensitive applications

Nomenclature

FFF	Fused filament fabrication
PLA	Polylactic acid
LW	Lightweight

S. Sahu (✉) · P. Kumar
Department of Mechanical and Industrial Engineering, Indian Institute of Technology Roorkee, Roorkee 247667, India
e-mail: sonikasahu12@gmail.com

N. Kumar et al. (eds.), *Advances in Materials and Agile Manufacturing*, Lecture Notes in Mechanical Engineering,
https://doi.org/10.1007/978-981-99-6601-1_9

1 Introduction

In recent years, the requirement for lightweight components has increased significantly in various weight-sensitive industries such as aerospace and automotive, and medical devices. In this, weight reduction can reduce energy consumption, can improve performance, and enhance patient comfort. With the technological advancements, nowadays polymers are replacing ferrous and non-ferrous metals in most of the weight-sensitive applications. Different strategies are adopted to fabricate lightweight components such as reinforced polymer composites, sandwich panels of polymers, and polymer metals hybrid. In order to create lightweight components, wide range of polymer materials is available. Manufacturing methods used to process these materials vary significantly. The selection of manufacturing method highly relies on the selected material and the desired geometry of the targeted components.

Due to the requirement of special tooling, conventional polymer manufacturing methods like injection and compression molding face difficulties in developing lightweight components having complex shapes. Nowadays, 3D printing has been emerged as a manufacturing method which has the potential to overcome these limitations [1, 2]. Among all the available 3D printing techniques, fused filament fabrication (FFF) is one of the most promising due to its affordability and simplicity. It is a popular and widely used 3D printing technology that involves the extrusion of molten material, typically plastic, layer by layer to create a 3D object [3, 4]. It offers various advantages over the conventional manufacturing methods such as capabilities to fabricate parts with any complexity, flexibility to vary internal structures [5–9].

It is known that high strength-to-weight ratio is very crucial factor in the weight-sensitive applications such as automobile, and aerospace. FFF has ability to develop lightweight components without affecting their performance and maintaining high strength-to-weight ratio.

In the literature, various research articles have been reported on the development of lightweight structures through FFF in which different approaches have been adopted such as design optimization, filler reinforcement, and sandwich panel. Therefore, efforts need to make toward the consolidation of literature in this area in order to identify the current status, gap and future scope. Thus, current article aims to review the existing articles related to the development of lightweight components using FFF. Various approaches have been presented with highlighting their advantages and disadvantages.

2 Fused Filament Fabrication

Fused filament fabrication (FFF) is a 3D printing process in which filament shape of material is processed to develop three-dimensional components. The filament is fed with the help of two co-rotating rolling wheels into the small opening of liquefier head. It works on the material extrusion principle in which filament itself acts as

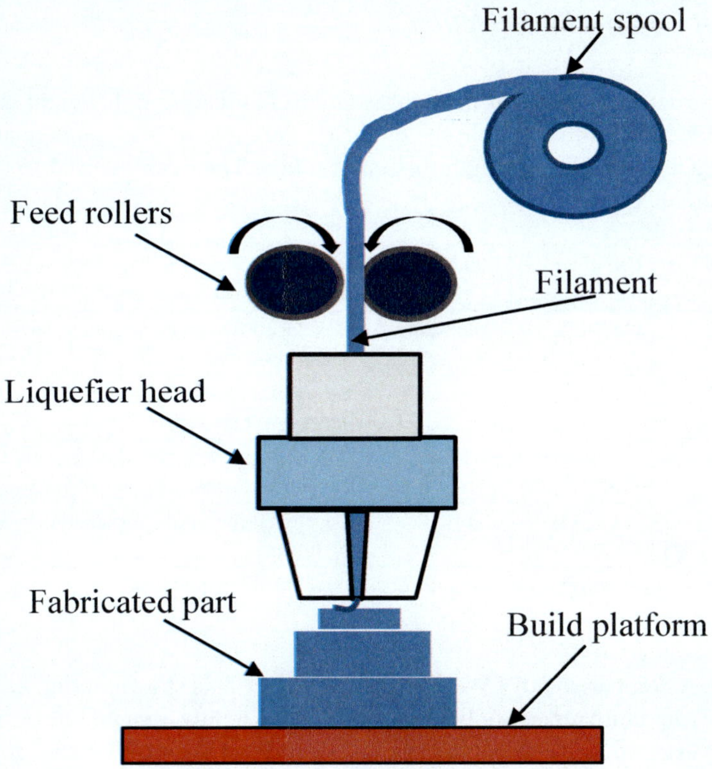

Fig. 1 Schematic diagram of FFF

a plunger and inner opening wall of liquefier head works as the cylinder as shown in Fig. 1. As a result, the material is extruded in thin form of wire through a tiny nozzle, placed at the bottom of liquefier head. The extruded material is deposited on the build platform in the predetermined toolpath pattern. The pattern and other factors may be selected based on the end-use applications of the components which are under consideration. Different softwares such as Cura and Slicer are available for processing CAD model and to perform slicing according to the selected factors and their levels.

3 Approaches Used for Fused Filament Fabrication of Lightweight Components

There are several approaches that can be applied to fabricate lightweight components using FFF. Using the following approaches, the components weight may be reduced without sacrificing strength and durability.

3.1 Material Selection

The selection of right material is key to fabricating lightweight components using FFF. Several lightweight materials are available in the commercial market for FFF which have high strength-to-weight ratios. They can be used to create strong and durable components.

S. No	Commercial name	Manufacturer	Remarks
1	LW-PLA	Colorfabb	Developed for model aircraft to have a lighter wing load and a lower stall speed
2	LW-ASA	ColorFabb	Developed for model aircraft to have a lighter wing load and a lower stall speed
3	VarioShore TPU	Colorfabb	Soft and flexible offer variable shore hardness, reduced weight, and density
4	ePLA-LW	eSun	Developed for model airplanes, drones, and cosplay costumes
5	PolyLight 1.0	3DLabPrint	Developed for model airplanes, drones

A comparison of LW-PLA and normal PLA is shown in Fig. 2.

Literature reports only few studies on the lightweight polylactic acid. Ansari et al. investigated the process parameters to fabricate low-cost lightweight parts using FFF [11]. The impact strength and hardness properties were studied of carbon fiber (CF) reinforced polylactic acid (PLA). Taguchi design of experiment technique was used to design the experimental runs. The performance of printed specimens was tested using Shore D durometer and impact tester. Results indicated the dependency of impact strength on print speed and infill density. Grid-type pattern yielded the maximum impact strength of 113.84 J/m at the highest temperature, medium print speed, and infill density. However, 37.95% enhancement was obtained in Shore D hardness than that of normal PLA. A set of optimized process parameters for lightweight parts was obtained as nozzle temperature 240 °C, print speed 120 mm/s, and infill density 50% with the grid type infill pattern. In another study, Özsoykal et al. evaluated the

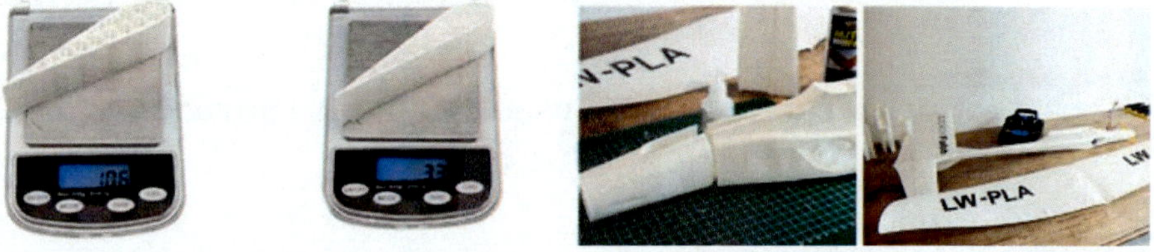

(a) (b)

Fig. 2 **a** Comparison of the weight of parts printed with standard PLA and LW-PLA **b** printed model of aircraft with LW-PLA. *Source* ColorFabb [10] radiological tissue equivalency of LW-PLA material printed using FFF [12]. They investigated the effect of process parameters such as printing temperature, flow rate, and infill density. The printed samples density was obtained between 0.63 and 1.19 g/cm^3 which was proportion to the infill density and flow rate of the material. However, average Hounsfield unit values of the printed samples obtained between $-$ 450 and $+$ 73 with the standard deviation range of $\pm$ 6 and $\pm$ 25.

3.2 Topology Optimization

Topology optimization uses evolutionary algorithms to identify the optimal shape and internal structure of the samples to reduce the weight without affecting their end-use functions, as shown in Figs. 3 and 4. It can help in fabricating the strong and durable lightweight components by redistribution of stress and eliminating unnecessary material.

Various studies have been reported on the topology optimization for additive manufacturing. The current and future trends of topology optimization were highlighted in the study carried out by Liu et al. [14]. Various algorithms were presented in their study about the support structure design, porous infill design, multi-material topology, etc. Primo et al. presented a case study of test specimen C-Clip in

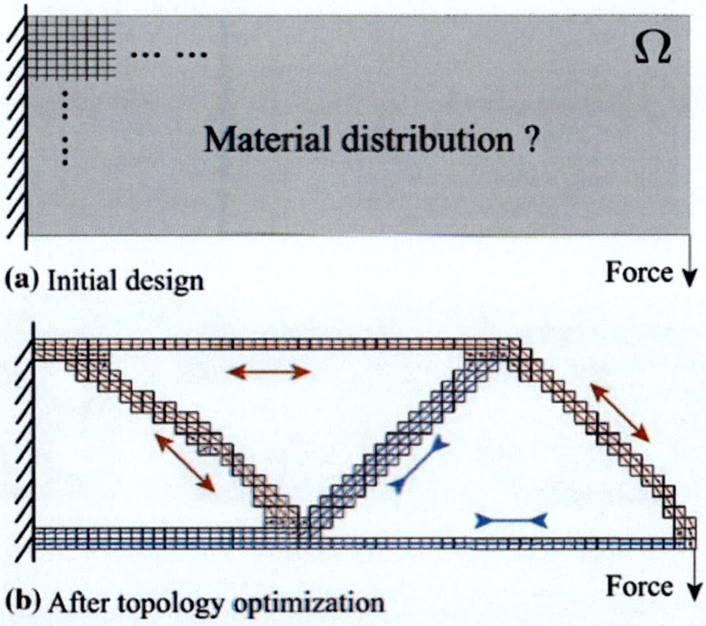

Fig. 3 **a** Redistribution of material as per load carrying direction **b** color code showing the principal stress directions: blue-compression and red-tension [13]. *Source* Liang et al.

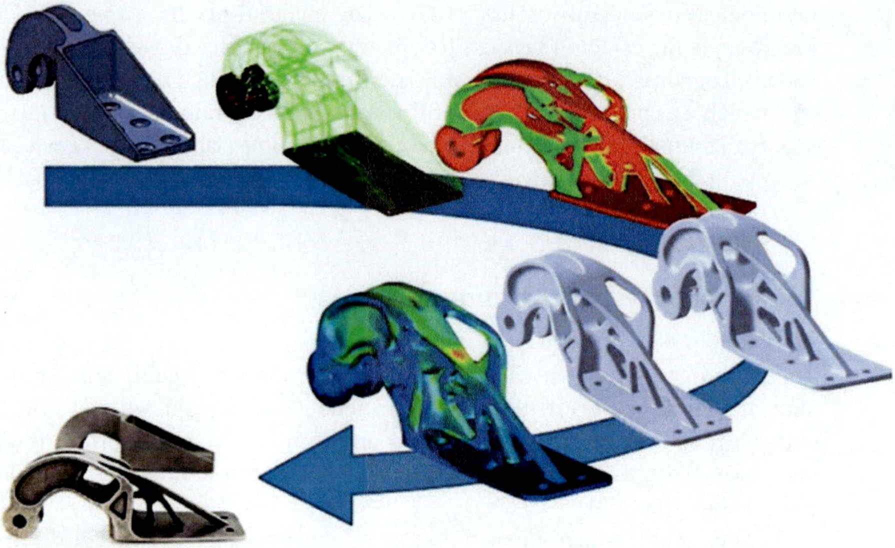

Fig. 4 Hinge bracket of airbus A320 nacelle redesigned through topology optimization [13]

which structural optimization was done using lattice design approach and topology optimization [15], as shown in Fig. 5.

The main aim of the study was to assess the advantages offered by the combined design approach of topology optimization and lattice structure. The appropriate key performance indicators (KPI) were developed in order to define an objective evaluation of the component performances. Post-processing of the obtained optimization results, an engineering intelligence tool was made. Finally, three best structural geometries were printed using 3D printing system.

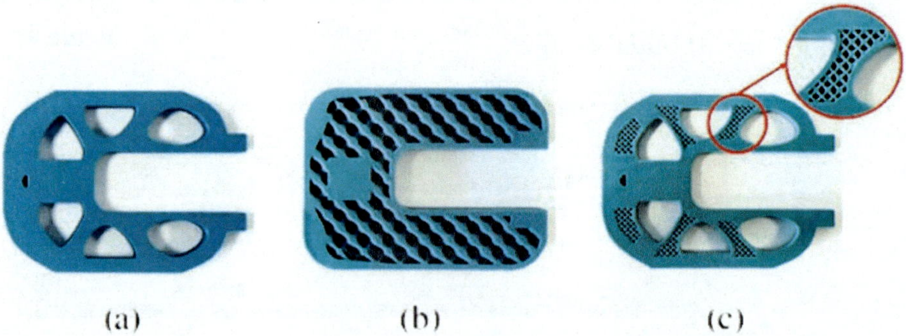

Fig. 5 Different structures of C-Clip obtained by: **a** topology optimization; **b** lattice structure; **c** hybrid approach [15]

3.3 Reinforced Polymer Composites

Material reinforcement is another approach adopted to develop lightweight components using FFF. It involves the use of additives such as fibers, nanoparticles, and microspheres that enhance the mechanical properties of the printed parts. These additives are mixed with the polymer matrix and then printed using FFF. Reinforcement involves the use of hollow microspherical particles such as glass bubbles and MWCNT [5, 16–23]. These composites are generally used in the weight-sensitive applications. Recently, this domain has attracted various researchers toward in the development of lightweight polymer filaments for the usage in weight-sensitive applications. Ashish et al. [24] blended hollow flyash cenospheres in HDPE and fabricated lightweight filament for commercial 3D printers to fabricate lightweight components. The fabricated filaments exhibited good mechanical properties than that of neat HDPE filament. Kumar et al. developed the polymer composites consisting high-density polyethylene (HDPE) and multiwalled carbon nanotubes (MWCNTs) [5]. Different weight percentage of MWCNT was considered in the study to realize the lightweight components. Thermogravimetric analysis (TGA) results showed the improvement in the thermal stability than that of neat HDPE at higher temperatures. Moreover, mechanical testing of the filament was carried out to obtain the change in tensile strength and flexural modulus which also indicated the improvement for the MWCNT-HDPE composites. In other studies, hollow particles of flyash cenospheres [20–22] were reinforced in the HDPE polymer matrix, as shown in Fig. 6. The lightweight filaments were developed by varying the volume of cenospheres. The developed filament material was characterized by investigating melt flow index and rheological properties. Cenospheres addition led to the enhancement in tensile modulus; however, strength was degraded. The developed material had the weight-saving potential to fabricate the components for weight-sensitive applications.

3.4 Part Consolidation

In most of assembly cases, there are always chances to consolidate multiple parts into a single component. It can also lead to weight reduction of the whole assembly by eliminating the redundant features and number of fasteners used. In the literature, some studies are found in which part consolidation approach has been used for weight reduction. Thomas Siebel presented an opinion research paper in which an example of seat mounting was considered to see the feasibility of the idea of part consolidation [25]. It was found that the additively manufactured seating mount was 40% lighter and 20% stiffer than that of the original one. Eight components of the seating mounting were replaced with one single part due to additive manufactuirng, as shown in Fig. 7.

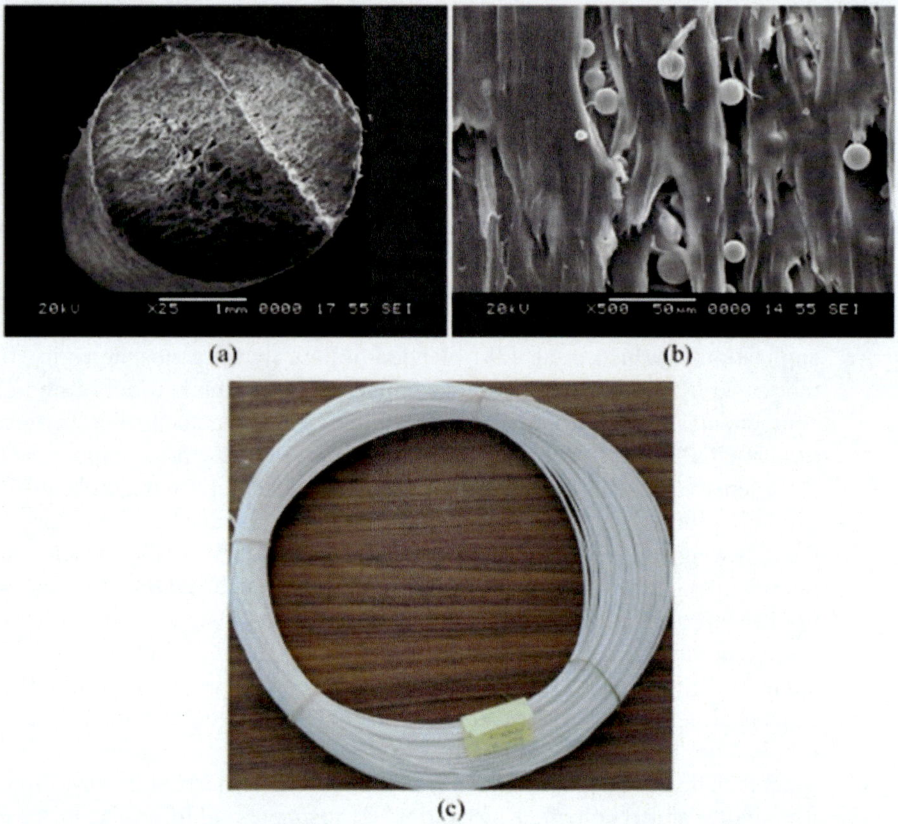

Fig. 6 Developed filament **a** knife-cut cross-section, **b** flysash presence in HDPE polymer matrix and **c** extruded filament [22]

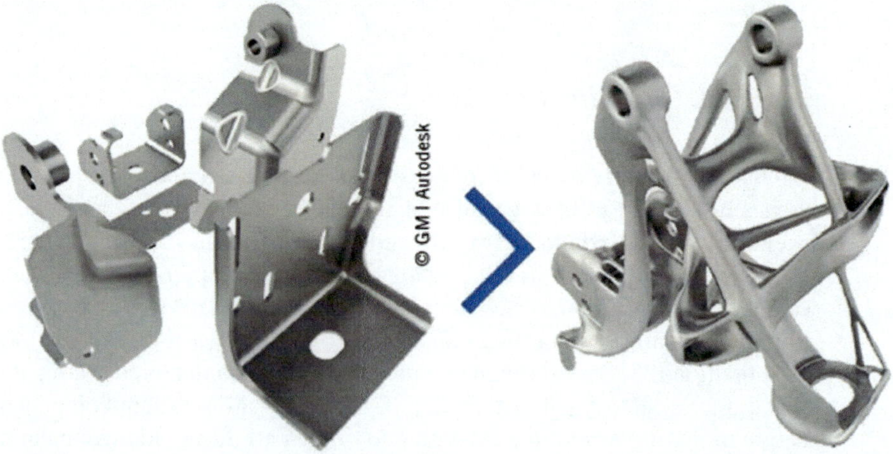

Fig. 7 Feasibility study of a seat mounting [25]

4 Conclusions

Fused filament fabrication 3D printing process has shown its potential in various applications. The current study presented different approaches which showed that the components made through this can also be in the weight-sensitive applications. The approaches such as lightweight material selection, topology optimization, reinforced polymer composites with hollow fillers, and part consolidation were presented with case studies. In the future, this study may be useful in various aerospace and automobile applications where lightweight components will be required. Electrical vehicles may also be targeted using FFF in the future as battery energy consumption is highly dependent on the overall weight.

Acknowledgement The authors express their gratitude to the Science & Engineering Research Board (SERB) for their generous financial support, which enabled the successful completion of this research (File Number: PDF/2021/002826).

References

1. Chua CK, Leong KF, Lim CS (2012) Rapid prototyping: principles and applications, 2nd edn, vol 2. World Scientific Publishing Co. Pte. Ltd., Singapore
2. Kumar N, Jain PK (2022) Extrusion-based additive manufacturing systems: current state, parameters optimisation, materials, research gap, challenges and future potential. Int J Manuf Technol Manag 36(1):28–64
3. Kumar A, Kumar N (2022) Advances in transparent polymer nanocomposites and their applications: a comprehensive review. Polym Technol Mater 61(9):937–974
4. Kumar A, Kumar N (2022) Parametric investigation into the transparency of PLA parts fabricated through fused filament fabrication. Adv Mater Process Technol 1–14
5. Kumar S, Ramesh MR, Doddamani M, Rangappa SM, Siengchin S (2022) Mechanical characterization of 3D printed MWCNTs/HDPE nanocomposites. Polym Test 114:107703
6. Singh S, Singh G, Prakash C, Ramakrishna S (2020) Current status and future directions of fused fi lament fabrication. J Manuf Process 55(April):288–306
7. Gawali SK, Kumar N, Jain PK (2020) Investigations on the development of heated build platform for additive manufacturing of large-size parts. In: Manufacturing engineering
8. Nath SD, Nilufar S (2020) An overview of additive manufacturing of polymers and associated composites. Polymers (Basel) 12(11):2719
9. Chadha A, Ul Haq MI, Raina A, Singh RR, Penumarti NB, Bishnoi MS (2019) Effect of fused deposition modelling process parameters on mechanical properties of 3D printed parts. World J Eng 16(4):550–559
10. Colorfabb LW-PLA. [Online]. Available: https://colorfabb.com/lw-pla-natural. Accessed 25 Jun 2023
11. Ansari AA, Kamil M (2022) Izod impact and hardness properties of 3D printed lightweight CF-reinforced PLA composites using design of experiment. Int J Light Mater Manuf 5(3):369–383
12. Özsoykal İ, Husemoglu RB, Yurt A (2022) Radiological evaluation of the effects of printing parameters on 3D printed cylindrical LW-PLA samples: preliminary results. J Med Innov Technol 3(2):28–34
13. Meng L et al (2020) From topology optimization design to additive manufacturing: today's success and tomorrow's roadmap. Arch Comput Methods Eng 27(3):805–830

14. Liu J et al (2018) Current and future trends in topology optimization for additive manufacturing. Struct Multidiscip Optim 57(6):2457–2483
15. Primo T, Calabrese M, Del Prete A, Anglani A (2017) Additive manufacturing integration with topology optimization methodology for innovative product design. Int J Adv Manuf Technol 93(1–4):467–479
16. Tambrallimath V et al (2022) Synthesis and characterization of flyash reinforced polymer composites developed by fused filament fabrication. J Mater Res Technol 21:810–826
17. Pejkowski Ł, Seyda J, Nowicki K, Mrozik D (2023) Mechanical performance of non-reinforced, carbon fiber reinforced and glass bubbles reinforced 3D printed PA12 polyamide. Polym Test 118(December 2022):1–11
18. Doddamani M (2020) Dynamic mechanical analysis of 3D printed eco-friendly lightweight composite. Compos Commun 19(March):177–181
19. Bharath HS, Sawardekar A, Waddar S, Jeyaraj P, Doddamani M (2020) Mechanical behavior of 3D printed syntactic foam composites. Compos Struct 254:112832
20. Bharath HS, Bonthu D, Prabhakar P, Doddamani M (2020) Three-dimensional printed lightweight composite foams. ACS Omega 5(35):22536–22550
21. Philosophy DOF, Patil B (2019) 3D printed syntactic foam
22. Bonthu D, Bharath HS, Gururaja S, Prabhakar P, Doddamani M (2020) 3D printing of syntactic foam cored sandwich composite. Compos Part C Open Access 3
23. Dizon JRC, Espera AH, Chen Q, Advincula RC (2018) Mechanical characterization of 3D-printed polymers. Addit Manuf 20:44–67
24. Singh AK, Patil B, Hoffmann N, Saltonstall B, Doddamani M, Gupta N (2018) Additive manufacturing of syntactic foams: part 1: development, properties, and recycling potential of filaments. JOM 70(3):303–309
25. Siebel T (2021) Additive manufacturing on the threshold of mass production. ATZ Electron Worldw 16(11):34–39

Monitoring Tool Wear with Sensors and Arduino in CNC Metal Cutting Turning Operation of Titanium Alloys (Ti6Al4V)

Sushil V. Ingle and D. N. Raut

Abstract CNC turning is a common operation done in industries. Tool with cutting condition at the chip tool interface is vital area for wear. To control and monitor tool wear, various methods and instruments have been developed for observing tool wear in actual operation mode. Collecting data while doing CNC turning operation was needed to understand, impact of process variables on tool wear and surface finish. This paper presents the importance of real-time data to monitor and control tool wear using sensors and Arduino. Specifically, sensor signals from CNC turning operations are highlighted and used to advance health checkups tools. Thus, real-time data will be useful to control process parameter which in turn will have impact on performance parameters.

Keywords Monitoring · Tool wear · Sensors · Arduino · Real-time data

1 Introduction

In metal turning operations, force value and computation are crucial because they relate to the design of machine parts, tool designs, power consumption's, vibrations, part precision, etc. The goal of measuring cutting force is to better understand how cutting occurs, including how cutting parameters affect cutting force, workpiece's machinability, creation of chips, chatter, and tool wear [1]. Research on cutting pressures in metal cutting operations has been conducted for more than century in an effort to better understand cutting behaviour. When compared to experimental results, it has been found that force levels derived from engineering computation often contain mistakes. Cutting operations are dimensionally difficult because under formed chip thickness and cutting velocity change constantly. Due to its intricacy, cutting forces are influenced by a wide range of factors even in steady-state settings,

S. V. Ingle (✉) · D. N. Raut
Department of Production Engineering, VJTI College of Engineering, University of Mumbai, Matunga (E), Mumbai, India
e-mail: ingle.sv@gmail.com

N. Kumar et al. (eds.), *Advances in Materials and Agile Manufacturing*, Lecture Notes in Mechanical Engineering,
https://doi.org/10.1007/978-981-99-6601-1_10

and the way that cutting forces change over time is distinctive. As a result, there are several reasons why measuring all cutting force components is necessary, but tool wear progress is perhaps the most crucial one [2].

Cutting force, which includes workpiece quality, cutting power, and tool wear, is a crucial indicator of the health of a metal cutting process [3–7]. Real-time cutting force measurements allow for the observation of cutting conditions and the timely adjustment of cutting parameters, all of which will increase cutting efficiency, prolong tool health, and save machining costs. High precise measurements of cutting pressure during maximum speed metal removing processes are becoming more and more necessary as machining techniques improve in terms of both accuracy and velocity. As a result, there is a considerable need for a force sensor with maximum accuracy and maximum natural frequency. Since the middle of the twentieth century, force calculation research has been ongoing for a considerable amount of time [8]; and numerous types of cutting force sensors, including present, vibrant, fibre-optic, strain gauge, and piezoelectric sensors, have been developed [9–12]. Limited of them meanwhile may meet demands of maximum precision and maximum natural frequency. As their great stability and effective performance, strain gauge and piezoelectric sensors have received much more attention to date. For instance, Ergun Ates and Kadir Aztekin created single piece strain gauge-based lathe sensor with sensing forces in two axes [13]. The sensor has a good measuring error (3.75%); however, it can only detect two of three components of force, while author was not able to explain how detected cutting forces interact with one another. Tulio Hallak Panzera created triple-axial pressure sensor having repeatability of lower than 8.4% using the same method. Unfortunately, his paper ignores natural frequency. Furthermore, since lathe tool post needs to be taken out in order to attach sensor, designed sensor was incompatible in lathe machine. Original lathe system may be impacted by this. Strain gauge kind of pressure sensor in turning was also developed by Yaldiz [14]. Portable piezoelectric sensor used in measuring triaxial pressure for turning was proposed by Totis and Sortino [15]. Finite element method calculation shows that work performs well, with static relative faults lower value to 5.91% and natural frequency being around one kHz. Smart tool for calculating force that was created by Chao Wang and colleagues has strong linearity and minimal cross-sensitivity [16]. Nevertheless, it is unable to measure constant force signals and only measure single component of force. In general, only a few number of cutting force sensors have the favourable precision and maximum natural frequency needed for measure forces under conditions of maximum precision and maximum speed metal cutting.

2 Sensors and Arduino

Smart sensor instrument assembly is a sensor device powered by a microprocessor that can gather three-dimensional vibrations. To accurately capture low amplitude signals and prevent any loading effects, signal measurement and amplification are conducted in the ADXL335 Triple-axis accelerometer circuit, which is depicted in

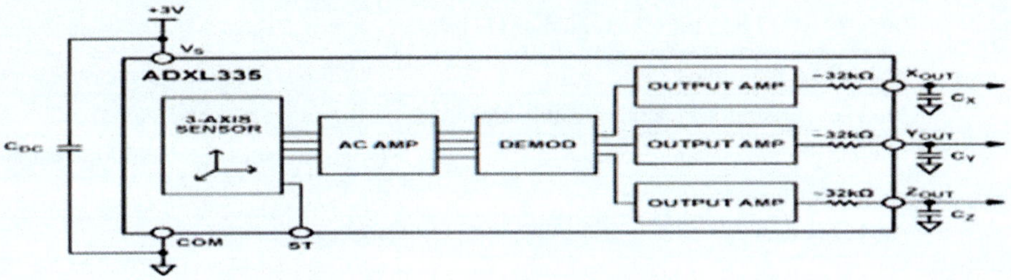

Fig. 1 Vibration measurement and amplification [17]

Fig. 1. The amplified signal is collected by the Arduino UNO microcontroller and delivered to any third-party device on a desktop or laptop computer via a serial communication protocol. Software for computers can be used to view the collected data (Tables 1 and 2) [17].

Table 1 Specification of Arduino UNO [17]

Microcontroller	AT-mega 328P
Operating voltage	5 V
Input voltage (recommended)	7–12 V
In–out voltage (limit)	6–20 V
Digital I/O pins	14 (6 provide PWM output)
PWM digital I/O pins	06
Analog input pins	06
Analog input pins	20 mA
DC current for 3.3 V pin	50 mA
Flash memory	32 KB (ATmega328P) of which 0.5 KB is used by the boot loader
SRAM	2 KB (AT-mega 328P)
EEPROM	1 KB (AT-mega 328P)
Clock speed	16 MHz
LED_BUILTIN	13
Length	68.6
Width	58.4
Weight	25 g

Table 2 ADXL335 triple-axis module [17]

The ADXL335 triple axis	Three-axis acceleration measurement
Module measures acceleration within the range	$\pm$ 3 g in the *x*, *y*, and *z*-axes
Module produces analog voltages	Proportional to acceleration as output signals
Highest voltage level	0 g is 1.65 V
Operating voltage	1.8–3.6 V
Operating current	350 μA
Temperature range	− 40 to + 85 °C
Sensitivity	270–330 mV/g
Shock resistance	Up to 10,000 g
Dimension	4 mm × 4 mm × 1.45 mm

3 Vibration Sensing and Amplification with ADXL335 Triple Axis

A compact, slim, less-power triaxial accelerometer sensor having signal-conditioned voltage outputs is ADXL335 Triple axis sensor. Gravitational constant, g, has a broad spectrum of approximately 3 g (1 g = 9.8 m/s^2). When using tilt sensors both stationary acceleration of gravity with robust acceleration brought on with movement, shock, or vibration can be detected it. The X_{out}, Y_{out}, and Z_{out} pin capacitors C_x, C_y, and C_z are used to control the bandwidth of accelerometer. Bandwidths for *X*-, *Y*-, and *Z*-axes can range from 0.5 to 550 Hz and from 0.5 to 1600 Hz, respectively. ADXL335 Triple-axis sensor comprises an open-loop acceleration measuring system using poly-silicon surface miniature machined sensor and data conditioning electronics. Output data analog voltages proportional to acceleration in tilt-sensing operations, an accelerometer may calculate both the stationary acceleration of gravity and dynamic acceleration brought on with movement, vibration, acquisition of vibration sensor, signal amplifier, and data [17].

Deflection of structure is calculated with inequitable capacitor with separate static plates and plates combined to rotating mass. Square waves that are 180° out of phase move stationary plates. Acceleration deflects a rotating mass, disrupting equilibrium of a differential capacitor, and creating sensor outcome whose amplitude is inversely proportionate to acceleration. Acceleration's amplitude with direction is determined using phase-sensitive demodulation algorithms (Table 3).

Material: The material used for the experiment is titanium alloy (Ti6Al4V) of 40 mm diameter and 80 mm long round bar. In all, there were four pieces of the same dimension and material for trial.

Tool details: Walter Insert DNMG110408 MP3 coated with PVD (TIALN) was used for experiment as per holder specification of the CNC MTAB MAXTURN machine. A total of 08 number inserts were used for 25 trials as per DOE.

Experimental Setup: CNC machine used for turning Ti6Al4V alloy is MTAB MAXTURN a small industrial turret lathe with eight programmable stations. FANUC

Table 3 Process parameters

Sr. No	Speed (Rpm)	Feed rate (mm/rev)	DOC (mm)
1	80	0.12	0.6
2	100	0.18	0.12
3	120	0.24	0.18
4	140	0.30	0.24
5	160	0.32	0.30
Symbol	*V*	*F*	*d*

MTAB is a machine that is available with commercial controllers. MAXTURN can be integrated with automation components such as a robot, and gantry for automating production. It is suitable for prototyping and batch production of light machining operations (Figs. 2, 3; Table 4).

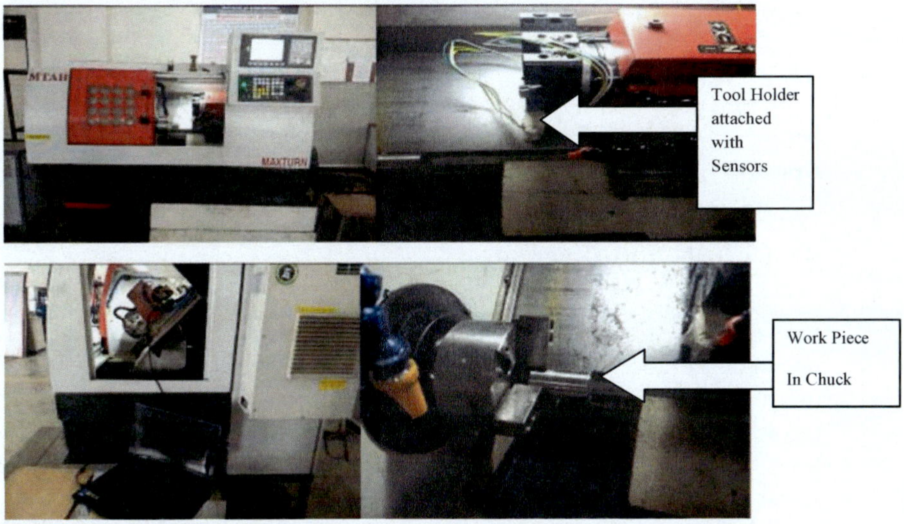

Fig. 2 CNC machine

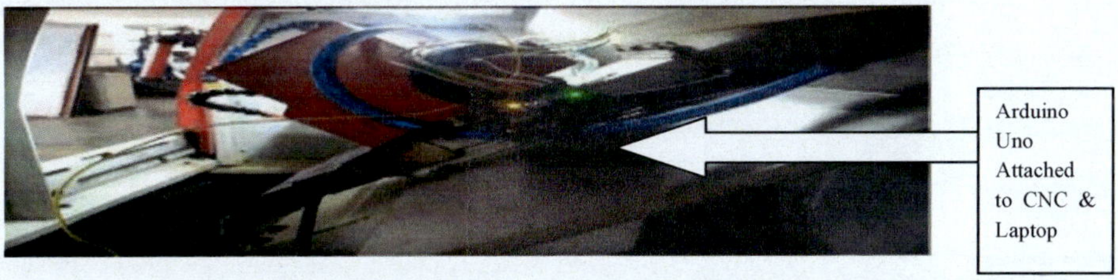

Fig. 3 CNC machine with Arduino

Table 4 Product specification

Model name/Number	MAXTURN
Max spindle speed	4000 rpm
Spindle motor power Siemens continuous	3.7 KW
Standard chuck size	135 mm
X-axis stroke	230 mm
Maximum turning diameter	80 mm
Maximum turning length	190 mm
Spindle size	A2-4

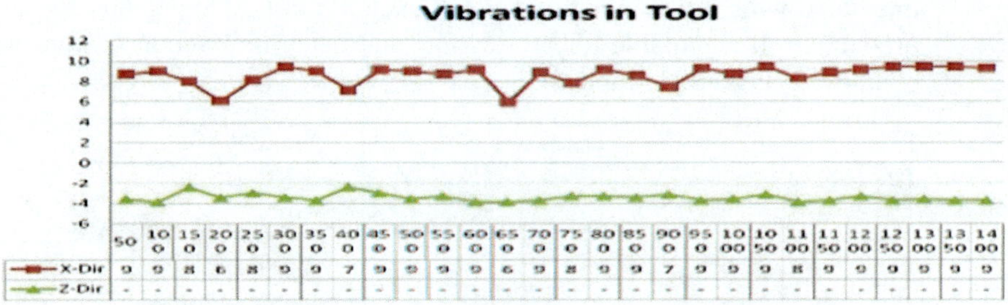

Graph 1 Vibration in tool

4 Results and Discussion

Real-time data for analysis is required or feedback is needed to control and monitor operation in processes. Real-time information helps to control sudden breakdown or damage by controlling machining parameters. Graph shows acceleration data received is plotted on X-axis against time (Sec). Data is acquired with the help of a sensor attached to tool holder during turning operation. During turning operation, sensor senses real-time data and sends it to Arduino UNO which transfers it with the help of software to the PC. The graph of acceleration data received is plotted on *Z*-axis against time (Sec).

Vibration in tool and tool holder reduces health of tool due to tool wear, chip formation, built-up edge, etc. By having real-time information about force, we can reduce source of vibration during turning operation hence improving surface roughness and better tool life (Graphs 1, 2, 3 and 4).

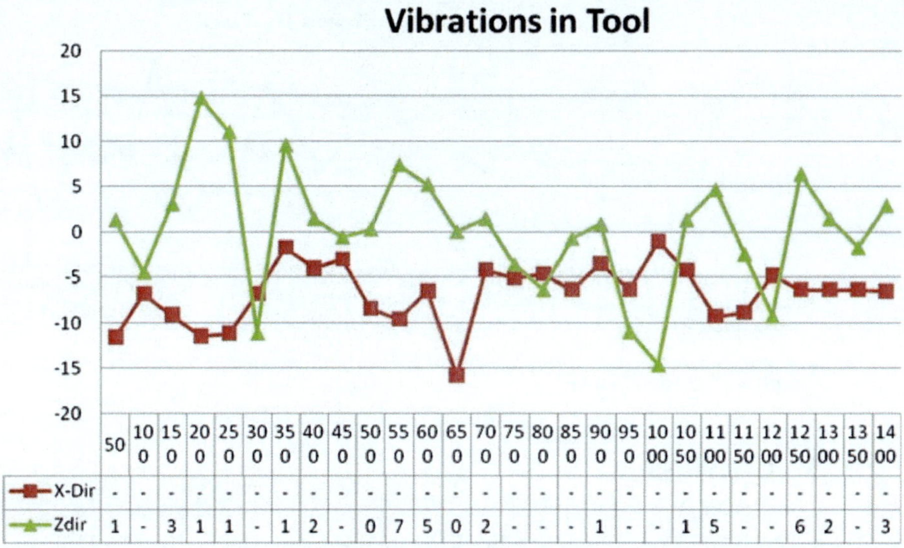

Graph 2 Vibration in tool

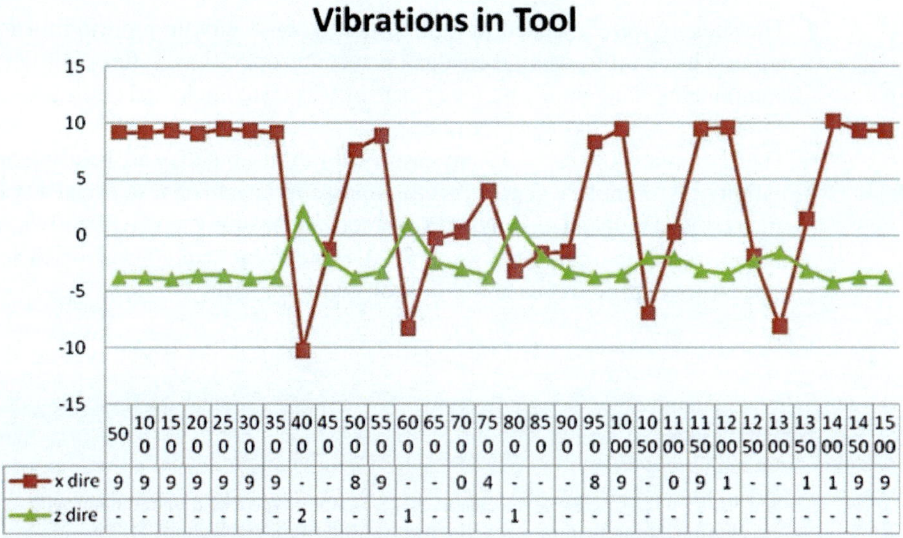

Graph 3 Vibration in tool

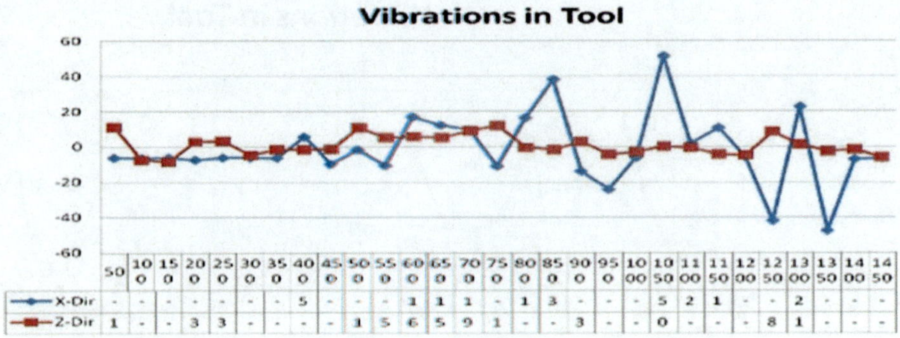

Graph 4 Vibration in tool

5 Conclusion

- While turning operation vibrations are observed which affect tool life and production quality and surface roughness. Hence to sense vibration on tool holder, the proposed approach to measure vibrations using a sensor attached to tool holder was successfully implemented.
- The sensors were attached to tool holder to sense vibration during turning operation. The reading was taken during turning operation. After estimating force components (tangential and feed), sensor was tested in actual cutting settings and demonstrated accurate performance.
- There are many instruments to measure vibrations, but they are costly. The combination of Arduino and sensors is easy to assemble, and easy to use at any location, and results are displayed on the PC screen, an Arduino can be directly attached to any PC with the help of software. So it's an economical alternative to accurately measuring cutting forces and vibrations with reliability.

References

1. Du KJ, Kim DS (1997) Development of a combined-type tool dynamometer with a piezo-film accelerometer for an ultra-precision lathe. J Mater Process Technol 71:360–366
2. Oraby SE-S (1989) Mathematical modelling and in-process monitoring techniques for cutting tools
3. Teti R, Jemielniak K, O'Donnell G, Dornfeld D (2010) Advanced monitoring of machining operations. CIRP Ann Manuf Technol 59:717–739
4. Dimla D (2000) Sensor signals for tool-wear monitoring in metal cutting operations—a review of methods. Int J Mach Tools Manuf 40:1073–1098
5. Abellan-Nebot JV, Romero SF (2010) A review of machining monitoring systems based on artificial intelligence process models. Int J Adv Manuf Technol 47:237–257
6. Sundara Murthy K, Rajendran I (2010) Design and development of strain gauge based milling tool dynamometer. Int J Mach Mach Mater 7:286–298

7. Rehorn AG, Jiang J, Orban PE (2005) State-of-the-art methods and results in tool condition monitoring: a review. Int J Adv Manuf Technol 26:693–710
8. Audy J (2006) Appraisal of techniques and equipment for cutting force measurement. J Zhejiang Univ Sci 7:1781–1789
9. Li X (2005) Development of current sensor for cutting force measurement in turning. IEEE Trans Instrum Meas 54:289–296
10. Spiewak SA (1995) Acceleration based indirect force measurement in metal cutting processes. Int J Mach Tools Manuf 35:1–17
11. Jin WL, Venuvinod PK, Wang X (1995) An optical fibre sensor based cutting force measuring device. Int J Mach Tools Manuf 35:877–883
12. Ri U, Wk WKH, Rq R, Dqg R (2012) Dinamómetro basado en sensores opticos de fibra Bragg. 4287–4290
13. Ateş E, Aztekin K (2013) Design, manufacturing, and calibration process of one piece lathe dynamometer for measurement in two axes. J Manuf Sci Eng 135:1–6
14. Yaldiz S, Ünsaçar F (2006) Design, development and testing of a turning dynamometer for cutting force measurement. Mater Des 27:839–846
15. Totis G, Sortino M (2011) Development of a modular dynamometer for triaxial cutting force measurement in turning. Int J Mach Tools Manuf 51:34–42
16. Wang C, Rakowski R, Cheng K (2013) Design and analysis of a piezoelectric film embedded smart cutting tool. Proc Inst Mech Eng, Part B: J Eng Manuf 227:254–260
17. Patel VK, Patel MN (2017) Development of smart sensing unit for vibration measurement by embedding accelerometer with the Arduino microcontroller. Int J Instrum Sci 2017:1–7

Design and Development of IoT-Based Controller for a Continuous Torrefaction Process

Shiv Prakash Dadhich, Kamal Kishore Khatri, Mohit Makkar, and Udayveer Singh

Abstract Torrefaction technology is applicable to convert raw biomass to valuable product charcoal. Torrefaction is preheated treatment process held at 200–300 °C with a residence time of 20–30 min in an inert atmosphere. To run the torrefaction process smoothly and remotely control the process parameters temperature and oxygen-free environment in a torrefaction reactor, an IoT-based RPi system is the best suitable approach. The proposed technology is an IoT-based RPi temperature control system, which improves the torrefaction process and optimizes process parameters like heating rate, temperature, and mass flow rate and reduces residence time to increase the production rate and quality of produced charcoal. In this paper, a k-type thermocouple is used to maintain a constant temperature and control the motor using RPi module. An HTML web page is developed to monitor process parameters and remotely control the torrefaction process. A PID controller is used to reduce the temperature and oxygen sensor's steady-state error and response time.

Keywords PID control · Raspberry Pi (RPi) · IoT system · Torrefaction · Proteus

Nomenclature

RPi	Raspberry Pi
IoT	Internet of Things
PID	Proportional Integral Derivative
PCB	Print Circuit Board
HTML	HyperText Markup Language

S. P. Dadhich (✉) · K. K. Khatri
Department of Mechanical and Mechatronics Engineering, LNMIIT, Jaipur 302031, India
e-mail: 19pmm003@lnmiit.ac.in

M. Makkar · U. Singh
Centre for Energy and Environment Studies, LNMIIT, Jaipur 302031, India

N. Kumar et al. (eds.), *Advances in Materials and Agile Manufacturing*, Lecture Notes in Mechanical Engineering,
https://doi.org/10.1007/978-981-99-6601-1_11

1 Introduction

World is currently facing energy crises due to the unavailability of fossil fuels. In village areas, most people burn agricultural waste, which is the main cause of environmental pollution. Torrefaction is the best suitable process to convert waste biomass to a useful product such as charcoal [1, 2]. The torrefaction process is the thermal treatment of biomass held at 200–300 °C temperature range with a residence time of 20–30 min and heating rate of 50 °C/min in an inert atmosphere [3, 4]. We can automate the torrefaction process to efficiently control process parameters temperature, inert atmosphere, and mass flow rate [By NTPC] [5]. Automation is the control of a process without the involvement of human beings [6]. We can control a process with many controllers like Arduino Uno with proper logic control [6]. PLC programming with ladder logic, Raspberry Pi, etc. [7]. A control system is mainly divided into two parts open control system and closed control system. A control without feedback is an open control system [8] (Fig. 1).

To remotely control process parameters of the torrefaction reactor, an IoT-based control system is best suited [9]. An HTML page is connected with an IP address through the RPi control system. We can change operating conditions based on sensor data. Torrefaction system is also modified with artificial intelligence using machine learning techniques [10]. The Raspberry control system can be used in two ways: coding in Python in mac software or creating a visual system in proteus software (Fig. 2).

IoT refers to data sharing through the internet between different energy domains like mechanical and electronic systems to control processes in remote locations [11–13]. The objective of IoT devices is to connect a physical system to a virtual mode and excess data all over the world using Internet. Nowadays, IoT devices are used

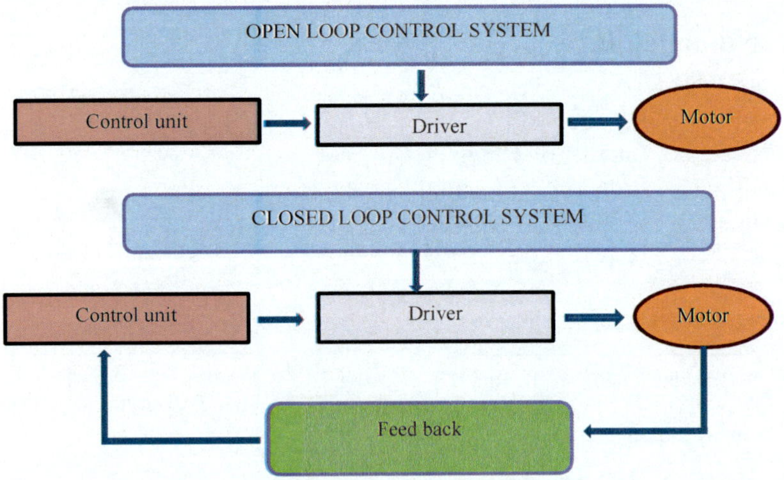

Fig. 1 Automation of a process

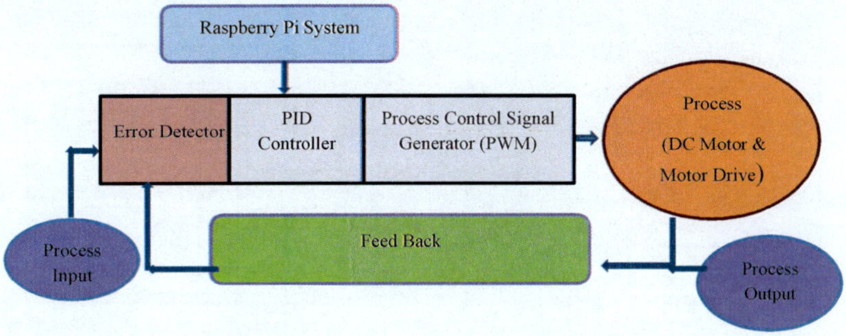

Fig. 2 Control process using Raspberry Pi

for many real-time applications like environment monitoring, traffic light control medical units, smart building and home automation, energy management, machine control [14–16]. IoT device selection depends upon the working machine. However, for simplicity and ease of operation and handling Arduino and Raspberry Pi are most suitable. Mostly GSM techniques are used for IoT applications as extensions, and RPi is a mini computer with RAM and memory [17–19].

In this paper, a continuous torrefaction IoT-based control system is designed and developed with using Raspberry Pi in proteus software. Steady-state error in the controller is optimized with PID Controller. An HTML web page is developed which is connected to the IP address of RPi for suitable action as per feedback data of process sensors.

2 Experimentation

2.1 Continuous Torrefaction Control System Design

Proteus software 8.1 is used for print circuit board (PCB) design to automate systems in different energy domains like mechanical and electronics. Simulation of the temperature control system proteus system 8.0 is used.

Sensors

The torrefaction process mainly depends on a reactor's temperature and inert atmosphere. To measure temperature in a reactor between temperature ranges 200–300 °C thermocouple is best suitable. We can control the temperature in a reactor using heat flow control at the LPG cylinder. To maintain an inert atmosphere in the reactor

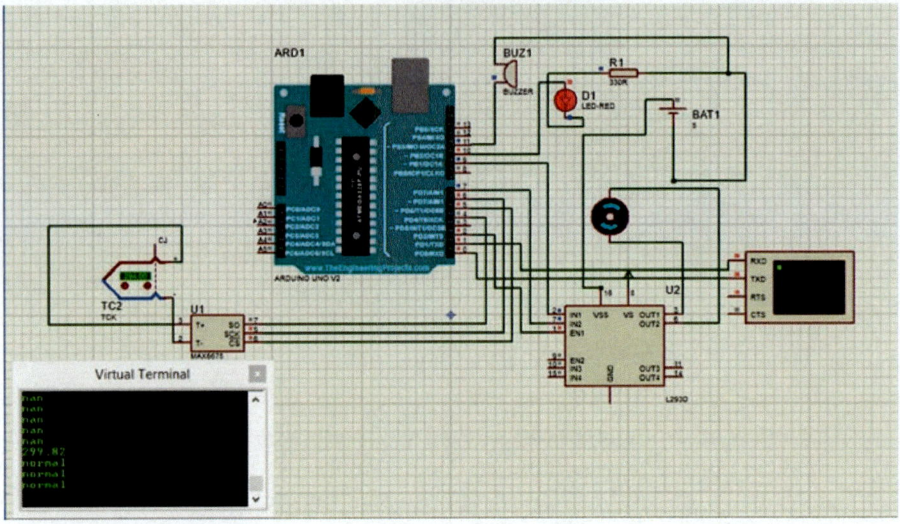

Fig. 3 Design of controller for continuous torrefaction reactor process using Arduino Uno

oxygen measurement sensor KN 25 is best suitable [20–22]. We can control the oxygen level in a reactor using nitrogen flow control at nitrogen cylinder.

A Design a controller using Arduino Uno

The proposed continuous torrefaction process control system consists of K type temperature sensor, a motor, an indicator, a buzzer, an Arduino, cloud, and visualization systems like mobile and desktop, L293D (Fig. 3).

To control the temperature in the desired range of 200–300 °C, a motor control system is used. When the system works in the desired range LED glows as an indicator. When the system is below the desired limit, the motor is allowed to move in the right direction; i.e., heat is supplied to the system to increase the temperature of system and buzzer starts with a beep sound as an indicator. When the system is above the desired limit, the motor is allowed to move in the left direction; i.e., heat is supplied to the system to increase the temperature of the system and buzzer starts with a beep sound as an indicator. We can develop a PCB board, attach a system with a torrefaction process, and control process at the plant.

B Design a controller using Raspberry Pi

To control a system with IoT applications on a web page first, we design it on proteus and then control it with IP address of the RPi minicomputer server. The proposed continuous torrefaction process control system consists of K type temperature sensor, a motor, an indicator, a buzzer, Raspberry Pi, cloud, and visualization systems like mobile and desktop (Fig. 4).

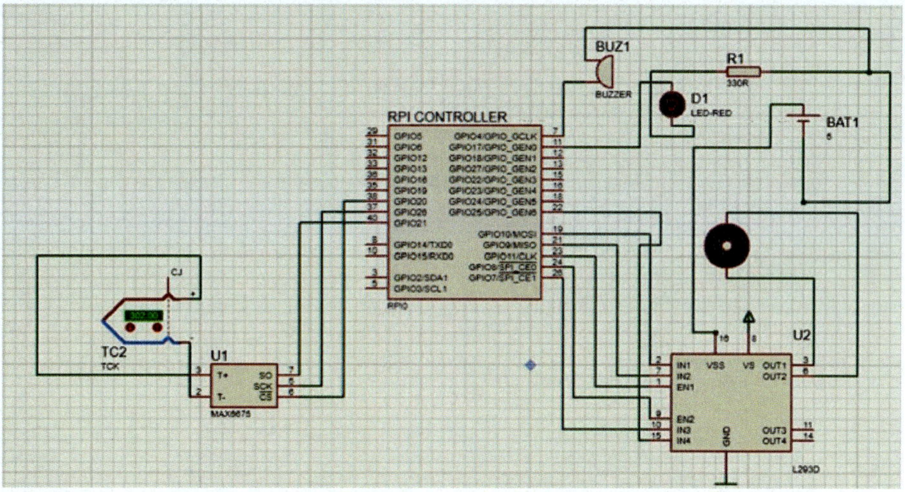

Fig. 4 Design of controller for continuous torrefaction reactor process using Raspberry Pi

2.2 *Methodology for Simulation and Validation of IoT-Based RPi Controller*

See Fig. 5.

2.3 *Experimental Setup*

Prepare a hardware electronic circuit using components like RPi, K-type thermocouple, buzzer, LED indicator.

A **Temperature control using Arduino Uno**

See Fig. 6.

B **Temperature and oxygen level control using Raspberry Pi**

To transfer data on a web server, RPi system is best suitable but there is only digital signal ports that are available with RPi to transfer signal and processing. Temperature and oxygen measurement sensors both are analog sensors. We add another module with RPi is analog-to-digital converter (ADC) is ADC1115 (Figs. 7 and 8).

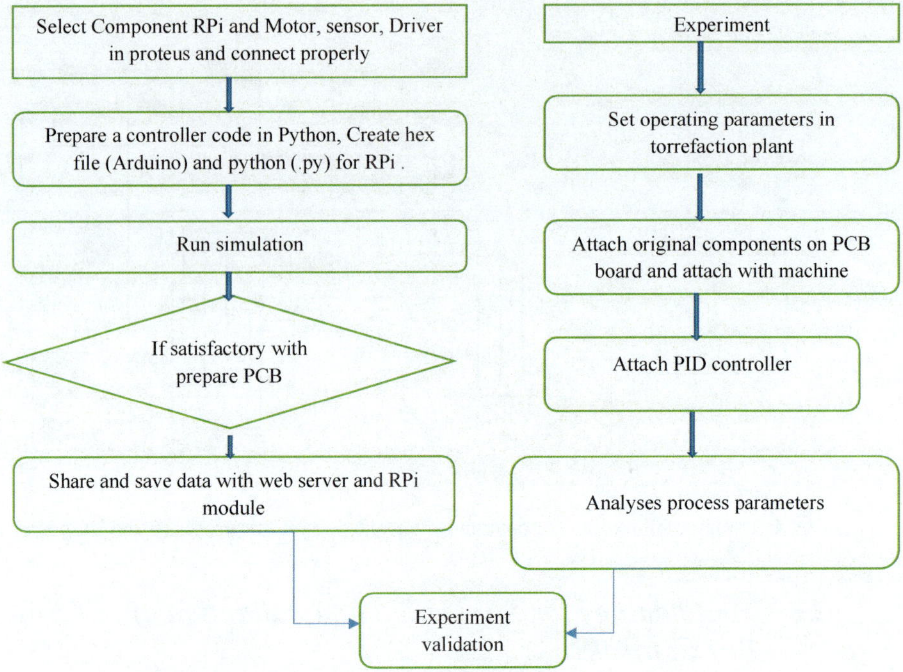

Fig. 5 Stages of RPi controller

Fig. 6 Temperature control using Arduino Uno

Fig. 7 Temperature control with RPi

2.4 *IoT Control System*

To share data on the cloud many platforms like web servers, ThingSpeak, Google Cloud, and ZigBee.

- **Android Mobile Application**

Many Android applications can turn ON/OFF signals and orient a switch in a desired output. In mobile applications, RaspController application is used to contact RPi GPIO pins using smartphones. Smartphones provide the facility to actuate the motors using an API key.

- **ThingSpeak Application**

To analyze data in real-world ThingsSpeak is open source. We can create a cloud on ThingSpeak for real setup simulation and machine data monitoring. ThingSpeak can coordinate with mobile applications and RPi to store and communicate data to an online server.

Fig. 8 Oxygen control with RPi

In the ThingSpeak channel, we can register using MATLAB Id and create a channel for temperature and inert atmosphere oxygen availability. Connect server to RPi using API key. We can store data in a web server for control using different techniques like machine learning.

- **HTML Web Server**

A Python code developed for HTML web page to control ON/OFF of the motor as per the desired temperature limit. All motors are connected with alphabet signals like A, a, B, b, and C, c for reference as set in RPi Python code (Fig. 9).

Motor 1: For temperature control, Motor 2: For inert atmosphere control.
Auto mode: For running the system smoothly conditions:
Manual Mode: For checking the system fault conditions.

3 Results and Discussions

The continuous torrefaction control system is designed in proteus software for simulation, results are validated using the ThingSpeak server, and system parameters are controlled on a web server.

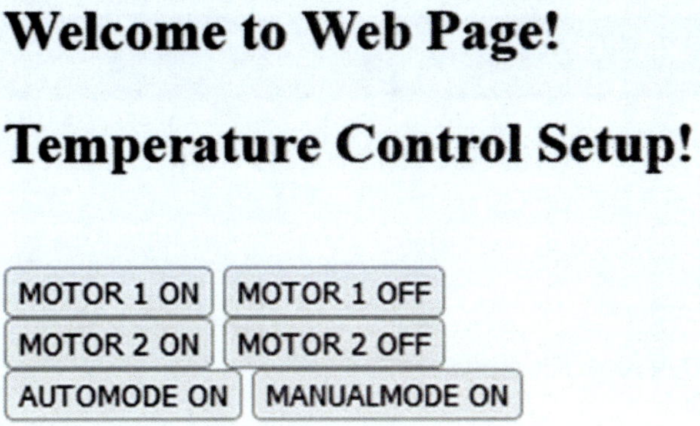

Fig. 9 Web page for torrefaction control process

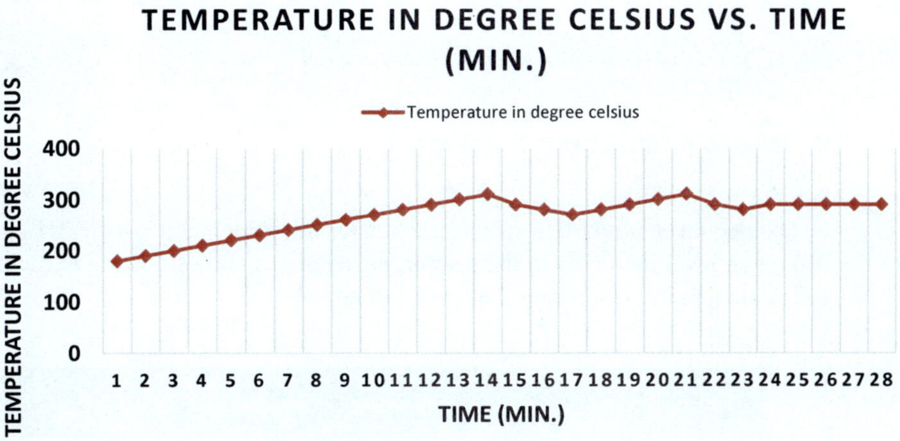

Fig. 10 Temperature control system using arduino uno

3.1 Temperature Control Using Arduino Uno

A heating zone for 200–300 °C is maintained in the torrefaction reactor with the control of heat flow in the reactor (Fig. 10).

3.2 Torrefaction Process Control System Using Rpi

A **Temperature Control using Raspberry Pi**

A heating zone for 200–300 °C is maintained in the Torrefaction reactor with the control of heat flow using on/off motor 1 as shown on a web page (Fig. 11).

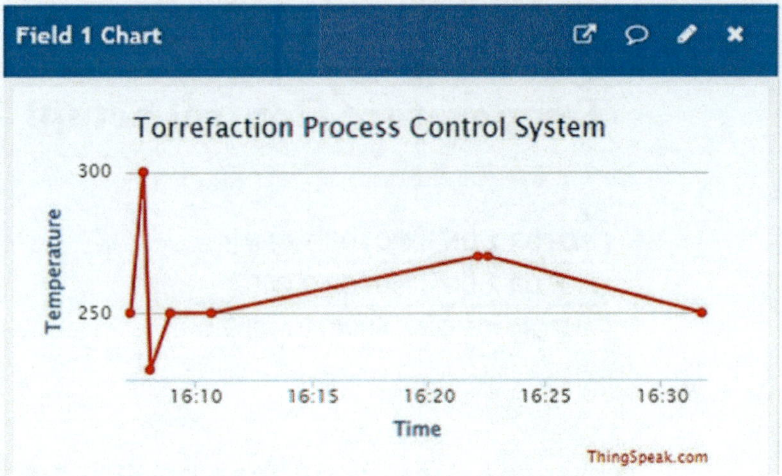

Fig. 11 IoT based temperature control system using Rpi

B Oxygen Control using Raspberry Pi

See Fig. 12.

Oxygen level required in the torrefaction reactor is maintained using KE25 sensor. If oxygen level increases in the reactor, nitrogen gas allows flowing to maintain an inert atmosphere; i.e., Motor 2 allows it to go on.

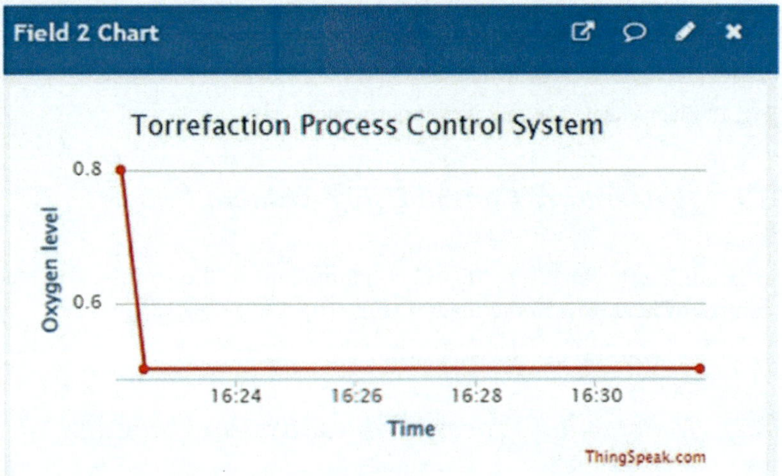

Fig. 12 Oxygen level control system

3.3 PID Controller for Process Parameter Optimization

PID controller is applied to reduce steady-state error such that the system can be optimized. Error signal passes to PID controller where K_p (Proportional gain), K_i (Integral gain), and K_d (Derivative gain) are varied to control the rotational speed of the reactor. To reduce steady-state error, Ziegler–Nichols method controls K_p, K_i, and K_d, where $e(t)$ = error function and $u(t)$ = PID controlled variable [16, 21–23] (Fig. 13; Tables 1 and 2).

Optimize PID Values

See Table 3.

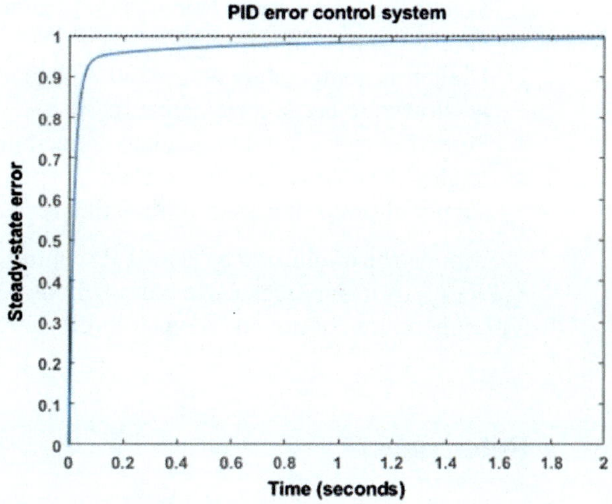

Fig. 13 PID control response

Table 1 Initial PID control variable

S. No.	Content	Value
1	Proportional gain (K_p)	1
2	Integral gain (K_i)	0
3	Capacity storage element (K_d)	0

Table 2 PID control response

S. No.	Particulars	Actual model	PID
1	Temperature	301	0300
2	Oxygen level (%)	0.99	1
3	Rise time (Sec)	0.024	0.024
5	Steady-state time (Rad/s)	0.3	0.1

Table 3 Final PID control variable values

S. No.	Content	Value
1	Proportional gain (K_p)	150
2	Integral Gain (K_i)	50
3	Capacity storage element (K_d)	30

4 Conclusions

IoT-based control and simulation of torrefaction reactor with process parameters temperature, inert atmosphere control are observed, and error is minimized with PID controller.

1. Lower the temperature below 200 °C motor 1 moves in right direction heat is allowed to pass in torrefaction reactor.
2. Higher the temperature below 300 °C motor 1 moves in left direction heat supply is allowed to cut in torrefaction reactor.
3. Potentiometer is used to regulate temperature ranges between 200 and 300 °C.
4. Higher the oxygen level more than 1% motor 2 moves in right direction nitrogen supply allowed to pass in torrefaction reactor.

Error can be minimized by using PID control system to maintain optimum value of K_p, K_i, K_d by using Ziegler–Nichols methods. Values are 150, 50, and 30 respectively for temperature control and oxygen level control system.

References

1. Strandberg M et al (2015) Effects of temperature and residence time on continuous torrefaction of spruce wood. Fuel Process Technol 134:387–398. https://doi.org/10.1016/j.fuproc.2015.02.021
2. Wang Z, Lim CJ, Grace JR (2019) Biomass torrefaction in a slot-rectangular spouted bed reactor. Particuology Chinese Soc Particuology 42:154–162. https://doi.org/10.1016/j.partic.2018.02.002
3. Michel J-B, McCormick M (2015) Experimental investigation of continuous torrefaction conditions of biomass residues for the subsequent use of torrefied pellets in domestic and district heating systems. In: 10th european conference on industrial furnaces and boilers. pp 1–22
4. Tumuluru JS et al (2011) Ehsan. Ind Biotechnol 7(October):384–401. https://doi.org/10.1089/ind.2011.0014
5. For I, Of E (no date) E.O.I from Indian START UPs torrefied pellet manufacturing plant for agri-waste expression of interest (EOI) from Indian START-UPs for "torrefied pellet manufacturing plant for agri-waste" (invitation for expression of interest). pp 1–12
6. Wilk M, Magdziarz A, Kalemba I (2015) Characterisation of renewable fuels' torrefaction process with different instrumental techniques. Energy 87:259–269. https://doi.org/10.1016/j.energy.2015.04.073
7. Subramanian KR (2018) Can automation eliminate human intervention? Int J Eng Manage Res (3). Available at: www.ijemr.net
8. Eslami A, Oxendine A (2008) AC 2008-607: A PLC project in a control course laboratory a PLC project in a control course laboratory

9. Gokhale AA (1989) Logical circuits and their applications 3(3):106–109. https://doi.org/10.12691/automation-3-3-13
10. Vishwakarma SK et al (2019) Smart energy efficient home automation system using IoT. In: Proceedings—2019 4th international conference on Internet of Things: smart innovation and usages, IoT-SIU 2019. IEEE, pp 1–4. https://doi.org/10.1109/IoT-SIU.2019.8777607
11. Danita M et al (2019) IoT based automated greenhouse monitoring system. In: Proceedings of the 2nd international conference on intelligent computing and control systems, ICICCS 2018. IEEE, (June 2018), pp 1933–1937. https://doi.org/10.1109/ICCONS.2018.8662911
12. Kelemen M et al (2015) Experimental identification of linear actuator properties. Acta Mechanica Slov 19(1):42–47. https://doi.org/10.21496/ams.2015.005
13. Virgala I et al (2017) Microcontroller for mechatronic systems resistive sensors. 7–12
14. David N et al (2015) Design-of-a-home-automation-system-using-Arduino.docx
15. Sukop M et al (2011) Increasing degree of automation of production systems based on intelligent manipulation. Acta Mechanica Slov 15(4):58–63. https://doi.org/10.21496/ams.2011.041
16. Zhao JC et al (2010) The study and application of the IOT technology in agriculture. In: Proceedings—2010 3rd IEEE international conference on computer science and information technology, ICCSIT 2010, vol 2. pp 462–465. https://doi.org/10.1109/ICCSIT.2010.5565120
17. Indra WA et al (2018) GSM-based smart energy meter with Arduino Uno. Int J Appl Eng Res 13(6):3948–3953. Available at: http://www.ripublication.com
18. Bhavani R, Alagammal S (2016) Design and implementation of GSM based smart energy meter (SEM) for home applications. Int J Latest Trends Eng Technol 8(1):431–439. https://doi.org/10.21172/1.81.055
19. Palanisamy A (2020) Home automation using PLC and Arduino. Acta Mechanica Slov 24(1):56–68. https://doi.org/10.21496/ams.2020.021
20. Jaiswal SP et al (2019) Internet of things (IoT) for smart agriculture and farming in developing nations. Int J Sci Technol Res 8(12):1049–1056
21. Gopi Krishna Rao PV, MVSKS (2013) Model based tuning of PID controller\n. J Control Instrum 4(1):16–22. Available at: http://www.stmjournals.com/index.php?journal=JoCI&page=article&op=view&path%5B%5D=3291%5Cn
22. Zhu Y, Song J, Dong F (2011) Applications of wireless sensor network in the agriculture environment monitoring. Procedia Eng 16:608–614. https://doi.org/10.1016/j.proeng.2011.08.1131
23. Lee M, Shamsuzzoha M, Vu TNL (2008) IMC-PID approach: an effective way to get an analytical design of robust PID controller. In: International conference on control, automation and systems, ICCAS 2008. pp 2861–2866. https://doi.org/10.1109/ICCAS.2008.4694246

Fabrication and Characterization of PLA Composite Filament for Medical Applications

Jasvir Singh, Vishal Francis, and Narendra Kumar

Abstract Additive manufacturing (AM) has demonstrated its potential in fabricating intricate and customized anatomical parts for biomedical applications. However, the widespread commercial adoption of 3D printing in biomanufacturing is hindered by the lack of readily available biocompatible, biodegradable, and chemically inert biomaterials. Fused filament fabrication technology has emerged as a promising approach to address this limitation by enabling the use of various polymers and polymer-based composites in medical applications, particularly for the fabrication of scaffolds, clinical devices, and tissue engineering systems. This research manuscript focuses on the development of a novel polylactic acid (PLA)-based biomaterial composite. The composite filament is thoroughly characterized using scanning electron microscopy (SEM), thermogravimetric analysis (TGA), and X-ray diffraction (XRD) techniques. These characterizations provide valuable insights into the structural and thermal properties of the developed filament, facilitating its potential application in biomanufacturing and biomedical engineering.

Keywords Additive manufacturing · 3D printing · Medical · PLA · Composite filament

1 Introduction

Additive manufacturing (AM) technology enables the direct fabrication of highly complex three-dimensional physical models and prototype parts using computer-aided design (CAD) data [1]. This layer-by-layer deposition technique, introduced in the early 1980s, has significantly enhanced the precision and efficiency of producing

J. Singh · V. Francis (✉)
School of Mechanical Engineering, Lovely Professional University, Phagwara , Punjab, India
e-mail: vishal.24813@lpu.co.in

N. Kumar
Department of Industrial and Production Engineering, Dr B R Ambedkar National Institute of Technology, Jalandhar, India

N. Kumar et al. (eds.), *Advances in Materials and Agile Manufacturing*,
Lecture Notes in Mechanical Engineering,
https://doi.org/10.1007/978-981-99-6601-1_12

3D prototypes compared to conventional manufacturing methods [2]. AM's potential extends beyond prototyping and offers various opportunities for advancement [3]. The initial drive behind the development of rapid manufacturing processes, like AM, was the desire to produce physical components quickly without relying on traditional tooling [4]. The goal was to reduce production lead times and enable the creation of more intricate geometric designs, which AM technology is now capable of achieving [5].

The significance of AM lies in its ability to facilitate effective communication between manufacturers and customers, reduce product development time, minimize errors, promote sustainable design practices, and provide opportunities for component replacement [6]. Fused deposition modeling (FDM) is a widely used AM technology, particularly in the field of biomedical applications [7]. FDM allows for the fabrication of complex geometries, customization of medical devices and implants and offers design flexibility [8]. Biocompatible materials, such as polylactic acid (PLA), can be utilized in FDM, making it suitable for producing biomedical devices that interact with the human body [9]. FDM is a promising and extensively utilized 3D printing technology, known for its ease of use, unsupervised operation, environmentally safe materials, and cost-effectiveness [10].

The FDM process involves the controlled melting of a polymer filament, extruding it through a nozzle, and depositing it layer by layer on a build platform [11]. To ensure proper material flow and layer bonding, the extruder head and build platform temperatures are regulated [12]. Support material is employed for overhanging features and can be removed either through physical breakage or chemical dissolution [13]. The density of the final part can be adjusted by modifying the infill ratio during the slicing process.

Although FDM using the OEM-based ABS P-400 filament from Stratasys Inc. is the common practice, its high cost poses challenges for commercialization in small- and medium-scale industries. Researchers have explored alternative feedstock filaments for FDM, including different polymers, abrasives, metal powders, and ceramics, to address this limitation [13–18]. The FDM process involves five steps: creating a CAD model, converting it to STL format, slicing the STL file into thin cross-sectional layers, constructing the model layer by layer, and finally cleaning and finishing the printed model [17]. AM technology has wide-ranging applications in the production of metallic, polymer, and composite parts with complex geometries, offering design flexibility while minimizing waste [19]. Its utilization spans various sectors, including biomedical, automotive, and aerospace industries [20]. FDM has gained significant traction in recent years and is expected to revolutionize manufacturing practices in numerous fields [21–23] (Fig. 1).

Pure polylactic acid (PLA) stands as the predominant printing media in the extrusion-based 3D printing technique owing to its capacity for decomposition. Yet, the unrestricted application of pure PLA polymer within the FDM strategy is restricted due to its limited mechanical performance [25, 26]. As a result, the incorporation of suitable additives is a decent strategy to improve the performance of pure PLA polymer for FDM [27, 28].

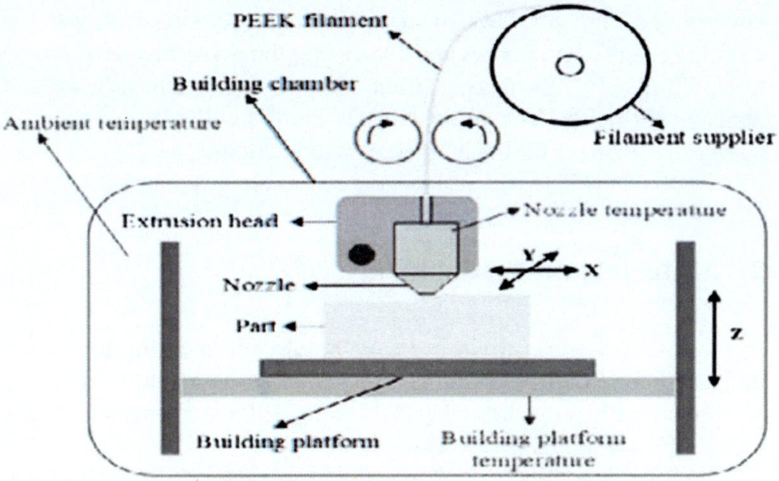

Fig. 1 Schematic diagram of FDM printer [24]

PLA is commonly employed for the fabrication of biomedical apparatus via FDM technique. PLA stands as a notably encouraging biodegradable polymer, given its economical nature for biomedical applications [29]. The unique properties of pure PLA further position it as a prime option for the fabrication of implant devices [30].

PLA can be produced from replenishable sources and undergoes biodegradation under elevated temperatures and humidity when subjected to composting [31–33]. It also possesses excellent processability which further adds value to it [27, 34–38]. Consequently, PLA finds application in various fields.

PLA is highly favored in biomedical applications due to its ability to meet diverse requirements, including biocompatibility, biodegradability, mechanical strength, and processability. To explore the broader range of biomedical applications for pure PLA, researchers have utilized PLA blends [39] with other polymers to enhance its properties or create unique properties tailored for specific applications. Various methods have been employed to address the limitations of PLA's physical properties or customize them for target applications. Common techniques for modifying PLA in bulk include copolymerization, crosslinking, adjusting composition, and blending [40].

The investigation of polylactic acid (PLA)-based composite materials with improved properties holds significant importance for biomedical applications. PLA is widely utilized in the biomedical field due to its biodegradability, biocompatibility, low cost, and ease of processing. However, pure PLA has certain limitations, such as mechanical weakness and water solubility, which may hinder its suitability for specific biomedical applications. Therefore, the development of PLA composites by incorporating suitable additives has emerged as a feasible approach to enhance the properties of 3D-printed PLA parts obtained through techniques like fused deposition modeling (FDM). These composite materials can offer improved mechanical strength, durability, and specific functionalities required for various biomedical devices and implants. Investigating PLA-based composites opens up avenues for tailoring material properties and optimizing the performance of biomedical applications, ultimately contributing to the advancement of healthcare technologies and patient well-being. The present paper discusses the development of PLA-based composite filament for the biomedical applications.

2 Material and Methodology

In the research work, a virgin PLA granule an emerging high-performance thermoplastic for FDM-3D printing is selected as base matrix material. The PLA is reinforced with polyether ether ketone (PEEK), carbon fiber, and graphene. The composition of filament is as follows:

PLA (95%) + [PEEK + Carbon fiber − (Particle Size 100 μm) + Graphene] (as Fillers) (5%). Figure 2 illustrates the basic methodology adopted for the investigations.

In the development of composite filaments for 3D printing in biomedical applications, a mixture comprising 95% PLA with remaining containing PEEK, carbon fiber, and graphene has been investigated. This composite formulation combines the desirable properties of each component to create a filament with enhanced characteristics. PLA provides biodegradability and biocompatibility, while PEEK may contribute excellent mechanical properties and chemical resistance. The addition of carbon fiber reinforces the filament, improving its strength and stiffness. Graphene, known for its exceptional electrical and thermal conductivity, may further enhance the overall properties of the composite filament. By combining PLA, PEEK, carbon fiber, and graphene, this composite formulation offers a promising approach for the 3D printing of biomedical devices with improved mechanical strength, durability, and conductivity, opening up new possibilities for advanced applications in the biomedical field.

The methodology employed in this study involved the development of a novel biomaterial using a combination of virgin polylactic acid (PLA) granules, polyether ether ketone (PEEK), graphene, and carbon fiber for reinforcement. The weight

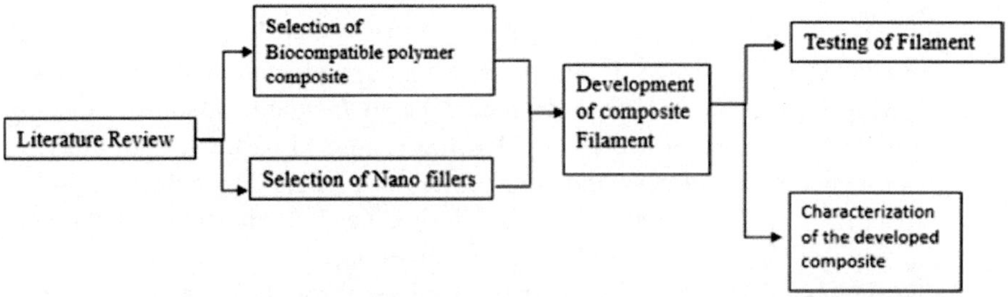

Fig. 2 Schematic of preparation process used to fabricate hybrid biopolymer filaments for 3D print

proportion of the components consisted of 95% PLA, along with PEEK, graphene, and carbon fiber. To begin, the PLA virgin granules were obtained and measured according to the desired weight proportion. PEEK, graphene, and carbon fiber were then added to the PLA granules, ensuring the appropriate weight ratios. The mixture of PLA, PEEK, graphene, and carbon fiber was homogenized using a double-cone blender to achieve a uniform distribution of the reinforcing materials within the PLA matrix. Next, the mixture was transferred to a chloroform solvent and placed on a magnetic stirrer. It was stirred for an adequate duration to allow for proper dissolution and homogenization of the components, resulting in a semi-solid viscous solution. The semi-solid viscous solution was then dried to form a solid casted thin layer of the composite material. It was spread evenly to obtain a uniform thickness and placed in an oven set at 60 °C to facilitate drying. The heat helped evaporate the chloroform solvent, leaving behind a solid composite material in the form of a thin layer. Subsequently, the solid casted thin layer of the composite material was cut into small flakes, which served as the raw material for the fabrication of the filament. An in-house developed filament extruder was used for this purpose. The small flakes were fed into the extruder, which processed and extruded them into a filament with a diameter of 1.75 mm (Fig. 3). Finally, the obtained filament underwent further characterization for subsequent investigations and analysis.

Figure 4 shows the line diagram of the filament extruder and the extruder parameters selected for the filament fabrication.

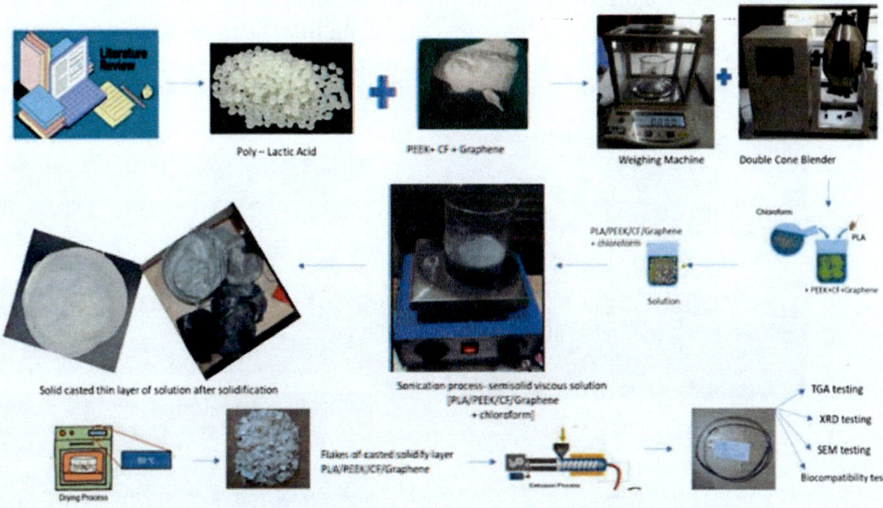

Fig. 3 Preparation process used to fabricate hybrid biopolymer filaments for 3D print

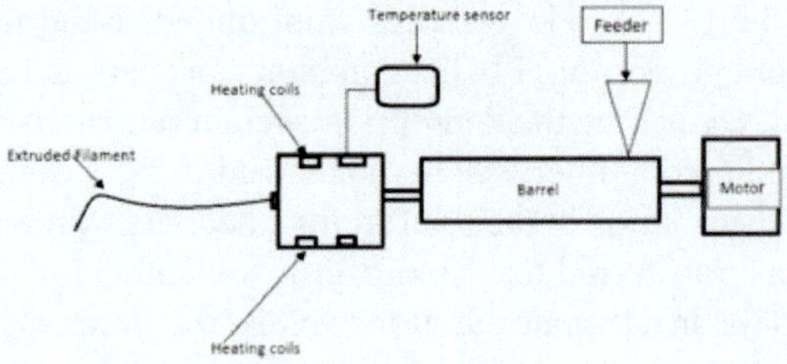

Parameter	Value
Motor speed	35 - 45 rpm
Operating heating coil temperature	190 °C - 200°C
Stirring	Magnetic stirrer at 60 rpm
Material Feed rate	at constant rate 10 gm/min

Fig. 4 Line diagram and parameters values of single screw extruder

3 Results and Discussion

Scanning electron microscope (SEM) analysis was conducted to investigate the microstructure and morphology of the composite filament. Figure 5 presents the SEM images of the PLA (95%) composite filled with PEEK, carbon fiber (particle size 100 μm), and graphene (5%).

The SEM images reveal a well-dispersed and uniform distribution of the fillers within the PLA matrix. The surface of the filament shows a smooth texture with no apparent agglomeration or clustering of the fillers. This indicates that the manufacturing process and blending techniques employed were effective in achieving a homogeneous distribution of the reinforcing materials within the composite. The presence of PEEK, carbon fiber, and graphene fillers can be observed, enhancing the structural integrity and potential mechanical properties of the composite filament.

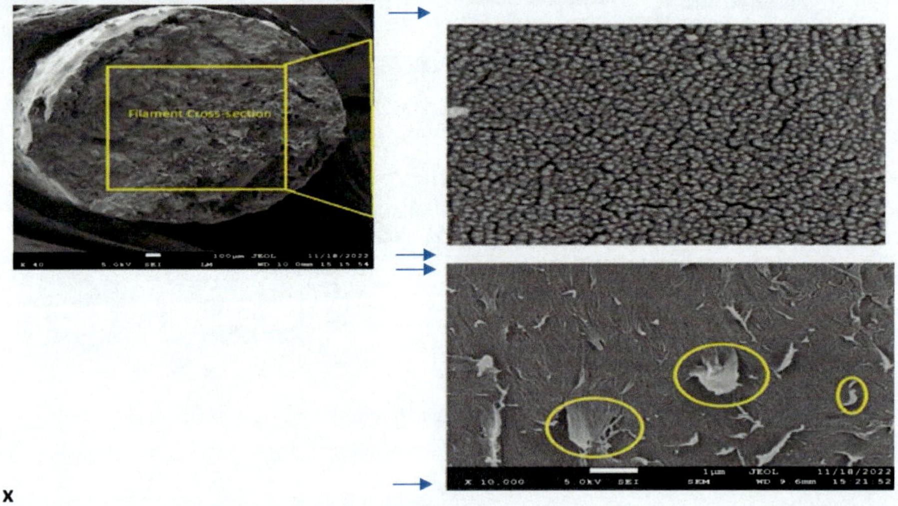

Fig. 5 SEM images of the developed composite filament

Thermogravimetric analysis (TGA) was performed to evaluate the thermal stability and decomposition behavior of the composite filament. The sample was subjected to a temperature ramp from 30.00 to 600.00 °C at a rate of 10.00 °C per minute.

The TGA results indicate that the composite exhibited thermal stability up to 235 °C. This means that the material remained stable and did not exhibit significant decomposition or weight loss up to this temperature. This enhanced thermal stability can be attributed to the incorporation of PEEK, carbon fiber, and graphene in the composite. These fillers likely contributed to the overall thermal resistance of the material, allowing it to withstand higher temperatures before undergoing significant degradation or decomposition. Figure 6 illustrates the thermal degradation pattern of the PEEK component under the influence of heat.

X-ray diffraction (XRD) analysis was carried out to examine the crystalline nature and phase composition of the developed composite filament. The XRD results reveal the presence of various elements in the composition, including PLA, PEEK, carbon fiber, and graphene. This confirms the successful incorporation of these fillers into the PLA matrix.

Moreover, the XRD analysis indicates a semicrystalline nature of the developed thermoplastic polymer composite. The diffraction patterns suggest the existence of both crystalline and amorphous regions within the composite material. This combination of crystallinity and amorphousness is favorable for achieving a balance between mechanical strength and flexibility in the composite structure. Figure 7 presents the XRD image of the PLA/PEEK/CF/graphene composite polymer, illustrating the diffraction peaks associated with the various elements present.

4 Conclusions

In conclusion, the utilization of PLA as the base material, along with PEEK, carbon fiber, and graphene as fillers, demonstrates the feasibility of fabricating a biopolymer composite filament. The comprehensive analysis of the composite filament yielded the following key findings:

The composite exhibited enhanced thermal stability, withstanding temperatures up to 235 °C, surpassing the thermal decomposition temperature of pure PLA. This indicates that the addition of PEEK, carbon fiber, and graphene contributed to the composite's ability to endure higher temperatures without significant degradation.

SEM analysis confirmed the uniform dispersion of PEEK, carbon fiber, and graphene within the PLA matrix. The low proportion of these fillers resulted in a homogeneous distribution throughout the composite filament, which is crucial for achieving optimal mechanical properties.

Overall, the developed PLA/PEEK/CF/graphene composite filament holds promise for biomedical applications, where a combination of biocompatibility, enhanced mechanical properties, and thermal stability is desired. Further investigations and characterization are warranted to explore the performance of this composite

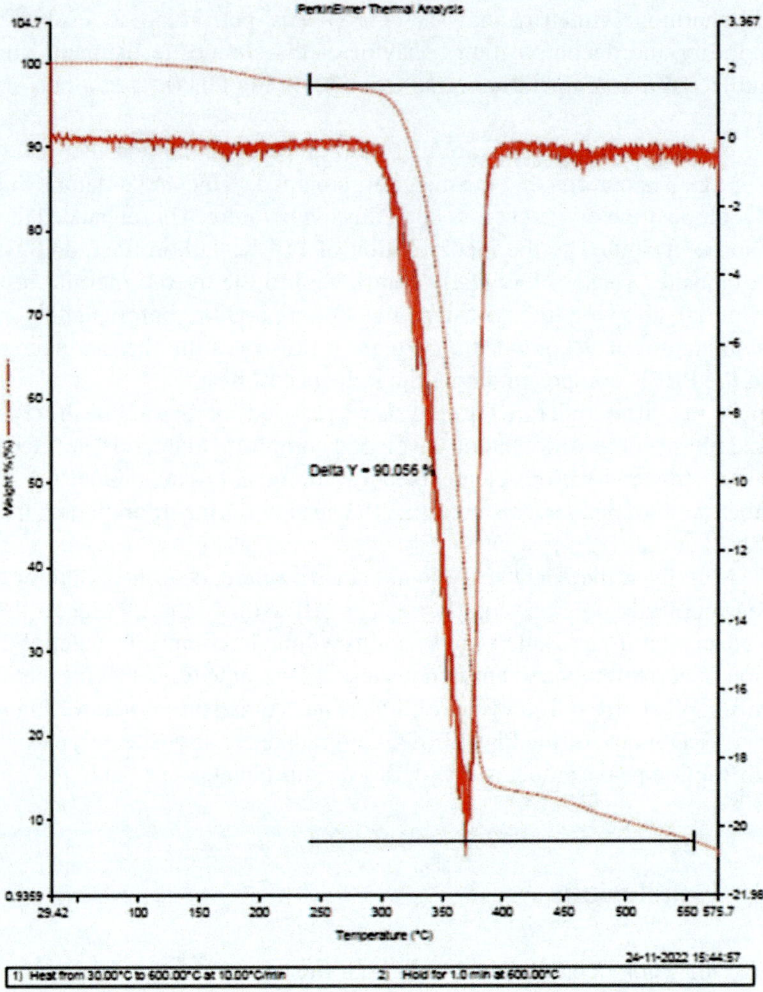

Fig. 6 TGA of developed composite filament

in specific biomedical applications such as scaffold fabrication and clinical device development.

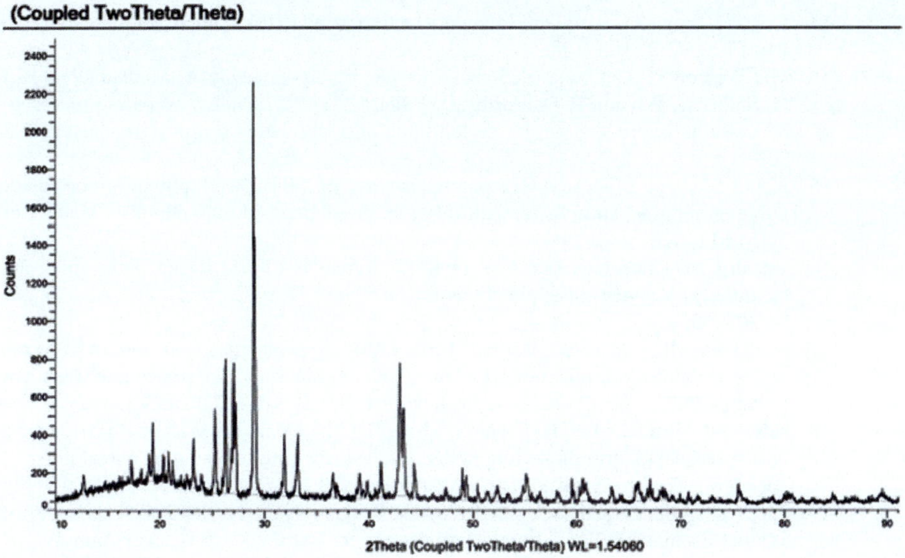

Fig. 7 XRD image of PLA/PEEK/CF/graphene composite polymer

References

1. Campbell I, Bourell D, Gibson I (2012) Additive manufacturing: rapid prototyping comes of age. Rapid Prototyping J 18(4):255–258
2. Negi S, Dhiman S, Sharma RK (2014) Basics and applications of rapid prototyping medical models. Rapid Prototyping J
3. Campbell RI, Dickens PM (1994) Rapid prototyping: a global view. Solid freeform fabrication: an advanced manufacturing approach. In: Marcus HL, Beaman JJ, Barlow JW, Bourell DL, Crawford RH (eds) 5th solid freeform fabrication symposium proceedings. Center for Materials Science and Eng., University of Texas, Austin, TX, pp 110–117
4. Milovanovic J, Trajanovic M (2007) Medical applications of rapid prototyping. Facta Univ Ser Mech Eng 5(1):79–85
5. Upcraft S, Fletcher R (2003) The rapid prototyping technologies, Emerals. Assembly Autom 23(4):318–330
6. Negi S, Dhiman S, Sharma RK (2014) Basics and applications of rapid prototyping medical models. Rapid Prototyping J 20(3):256–267. © Emerald Group Publishing Limited [ISSN 1355-2546]
7. Liua X, Chub PK, Ding C (2004) Surface modification of titanium, titanium alloys and related materials for biomedical applications. Mater Sci Eng R 47:49–121
8. Anitha R, Arunachalam S (2001) Critical parameters influencing the quality of prototypes in fused deposition modeling. J Mater Process Technol 113(1–3):385–388
9. Choudhury A, Chakraborty D, Reddy B (2007) Extruder path generation for curved layer fused deposition modeling. Comput-Aided Des J 40:235–243
10. Chhabra M, Singh R (2011) Rapid casting solutions a review. Rapid Prototyping J 17:328–350
11. Masood SH (1996) Intelligent rapid prototyping with fused deposition modelling. Rapid Prototyping J 2(1):24–32
12. Liu T, Wang M, Weizhi W, Zhang W (2008) Melt rheological properties of nylon 6/multi-walled carbon nano-tube composites. Compos Sci Technol 68:2498–2502

13. Singh R, Singh S (2014) Development of nylon based FDM filament for industrial applications. J Inst Eng 92(2):103–108
14. Van Weeren R, Agarwala M, Jamalabad VR, Bandyophadyay A, Vaidyanathan R, Langrana N, Safari A, Whalen P, Danforth SC, Ballard C (1995) Quality of parts processed by fused deposition. In: Proceedings of the solid freeform fabrication symposium, Austin, TX. pp 314–321
15. Kalita SJ, Bose S, Hosick HL, Bandyopadyay A (2003) Development of controlled porosity polymer-ceramic composite scaffolds via fused deposition modeling. Mater Sci Eng: C 20(5):611–620
16. Cheah CM, Chua CK, Lee CW, Feng C, Totong K (2005) Rapid prototyping and tooling techniques: a review of applications for rapid investment casting. Int J Adv Manuf Technol 25:308–320
17. Novakova ML, Novak MJ, Barna J, Torok J (2012) Special materials used in FDM rapid prototyping technology applications. In: IEEE 16th international conference intelligent engineering systems (INES), pp 73–76. 12–15th June, E-ISBN: 978-1-4673-2693-3
18. Mireles J, Kim HC, Lee H, Espaln I, Medna D, MacDonald F, Wicker R (2013) Development of a fused deposition modelling system for low melting temperature metal alloy. J Electron Packaging, Trans ASME 134(4). https://doi.org/10.1115/1.4007160
19. Tappa K, Jammalamadaka U (2018) Novel biomaterials used in medical 3D printing techniques. J Funct Biomater 9(1):17. https://digitalcommons.wustl.edu/open/access/pubs/6730
20. Torrado AR, Shemelya CM, English JD, Lin Y, Wicker RB, Roberson DA (2015) Characterizing the effect of additives to ABS on the mechanical property anisotropy of specimens fabricated by material extrusion 3D printing. Addit Manuf 6:16–29
21. Abourayana HM, Dobbyn PJ, Dowling DP (2018) Enhancing the mechanical performance of additive manufactured polymer components using atmospheric plasma pre-treatments. Plasma Process Polym
22. Horn TJ, Harrysson OLA (2012) Overview of current additive manufacturing technologies and selected applications. Sci Prog 95:255–282
23. Guo N, Leu MC (2013) Additive manufacturing: technology, applications and research needs. Front Mech Eng 8:215–243
24. Zhao F, Li D, Jin Z (2018) Preliminary investigation of poly-ether-ether-ketone based on fused deposition modeling for medical applications. Mat 11:288
25. Vaezi M, Seitz H, Yang S (2013) A review on 3D micro-additive manufacturing technologies. Int J Adv Manuf Technol 67:1721–1754
26. Turner BN, Strong R, Gold SA (2014) A review of melt extrusion additive manufacturing processes: I. Process design and modeling. Rapid Prototype J 20:192–204
27. Tümer EH, Erbil HY (2021) Extrusion-based 3D printing applications of PLA composites: a review. Coatings 11(4):390
28. Baran E, Erbil H (2019) Surface modification of 3D printed PLA objects by fused deposition modeling: a review. Colloids Interfaces 3:43
29. Parandoush P, Lin D (2017) A review on additive manufacturing of polymer-fiber composites. Compos Struct 182:36–53
30. Van den Eynde M, Van Puyvelde P (2018) 3D printing of poly (lactic acid). Adv Polym Sci 282:139–158
31. Wickramasinghe S, Do T, Tran P (2020) FDM-Based 3D printing of polymer and associated composite: a review on mechanical properties, defects and treatments. Polymers 12:1529
32. Rebelo R, Fernandes M, Fangueiro R (2017) Biopolymers in medical implants: a brief review. Procedia Eng 200:236–243
33. Santoro M, Shah SR, Walker JL, Mikos AG (2016) Poly(lactic acid) nanofibrous scaffolds for tissue engineering. Adv Drug Deliv Rev 107:206–212
34. Royte E (2006) Corn plastic to the rescue. Smithson Mag 37:84–88
35. Avinc O, Khoddami A (2009) Overview of poly(lactic acid) (PLA) fibre. Fibre Chem 41:391–401

36. Sawyer DJ (2003) Bioprocessing—no longer a field of dreams. In: Macromolecular symposia. Wiley, Hoboken, NJ, USA
37. Nofar M, Sacligil D, Carreau PJ, Kamal MR, Heuzey M-C (2019) Poly(lactic acid) blends: processing, properties and applications. Int J Biol Macromol 125:307–360
38. Scaffaro R, Lopresti F, Marino A, Nostro A (2018) Antimicrobial additives for poly(lactic acid) materials and their applications: current state and perspectives. Appl Microbiol Biotechnol 102:7739–7756
39. Saini P, Arora M, Ravi Kumar MNV (2016) Poly(lactic acid) blends in biomedical applications. Adv Drug Deliv Rev. https://doi.org/10.1016/j.addr.2016.06.014
40. Xiao L, Wang B, Yang G, Gauthier M (2012) Modification of PLA. In: Ghista DN (ed) Biomedical science, engineering and technology. InTech, Rijeka, Croatia, pp 255–259

Experimental Investigations on Biodegradable Polymer Fabricated Using Extrusion-Based Additive Manufacturing

Vishwajeet bhagat, Ratnesh Kumar, Jay Kumar, Sayantan Bhattacharya, Vishal Francis, Manjeet Singh, and Narendra Kumar

Abstract The utilization of additive manufacturing (AM) techniques has revolutionized modern manufacturing processes, allowing for the production of intricate geometries with enhanced design flexibility. Within this context, the exploration of biodegradable materials for AM applications has garnered significant attention due to their environmentally friendly characteristics and potential for sustainable manufacturing. This paper focuses on the experimental analysis of a biodegradable polymer, specifically polylactic acid (PLA), fabricated using extrusion-based AM. Tensile and compression testing, conducted in accordance with ASTM standards, provides valuable insights into the mechanical properties of the PLA components. Additionally, during the fabrication process of the specimens, two different infill orientations (0° and 90°) were employed and analyzed. By incorporating these distinct infill orientations, the study aimed to assess the influence of infill pattern orientation on the mechanical behavior of the fabricated specimens. The analysis of these different infill orientations provides valuable insights into the structural integrity and performance of the specimens under various loading conditions, further enhancing our understanding of the impact of infill orientation on the mechanical properties of the biodegradable polymer components. The experimental analysis contributes to a comprehensive evaluation of the mechanical behavior of the biodegradable polymer, supporting advancements in AM technologies and promoting sustainable manufacturing practices.

Keywords Additive manufacturing · 3D printing · PLA · Biodegradable

---

V. bhagat · R. Kumar · J. Kumar · S. Bhattacharya · V. Francis (✉) · M. Singh
School of Mechanical Engineering, Lovely Professional University, Punjab, India
e-mail: vishal.24813@lpu.co.in

N. Kumar
Department of Industrial and Production Engineering, Dr B R Ambedkar National Institute of Technology, Jalandhar, India

N. Kumar et al. (eds.), *Advances in Materials and Agile Manufacturing*,
Lecture Notes in Mechanical Engineering,
https://doi.org/10.1007/978-981-99-6601-1_13

1 Introduction

The process of additive manufacturing (AM) begins with the creation of a three-dimensional model using computer-aided design (CAD) software or by scanning an existing object [1]. AM is an established technology that has traditionally been used to make prototypes of all shapes, but it is increasingly gaining appeal in the manufacture of functioning parts. Applications for it are quickly spreading across various fields [2]. Any procedure that allows for the gradual addition of meticulously calculated two-dimensional cross-sectional layers to form a three-dimensional end product can be classified as AM [3]. In contrast to conventional subtractive manufacturing methods, AM excels at producing intricate or unique structures, including internal voids and complex features that are challenging to fabricate using traditional cutting tools [4]. AM has the capability to work with a diverse range of materials, including plastic, metal, and ceramics [5]. When it comes to polymer and composite materials, fused deposition modeling (FDM)/fused filament fabrication (FFF) is the preferred choice among various AM technologies. This is due to its flexibility, faster printing speed, affordability, high strength and toughness, non-toxicity, and wide variety of compatible materials. It outperforms other AM technologies in these aspects [6]. The FDM system operates on the principle of feeding filament into a liquefier using a motor, where it is melted. The molten material is then layer by layer deposited to create the completely formed portion. The simplicity and versatility of the FDM system have contributed to its popularity among users [7]. The flexibility of the FDM process enables the creation of diverse mesostructures, making it a straightforward solution to address this issue. By printing infill with varying densities and patterns, the optimal structure can be established [8]. Currently, the FDM process finds significant application in education and prototype assembly. However, as the technology continues to advance, its utilization is progressively expanding into diverse sectors including bioengineering, aerospace, and automotive industries [9].

Examining the mechanical characteristics of 3D printed components becomes crucial when considering the substitution of conventionally manufactured parts. Mechanical property tests offer valuable insights into material properties, such as tensile strength and compressive strength, facilitating a comprehensive understanding of the 3D printed parts [10]. The FDM printers have the capability to produce various pattern fills for the internal structure of printed parts, leading to different mechanical properties. However, a comprehensive definition and study of the characteristics of these printed pattern fills within the internal structure of the parts are yet to be fully explored [11]. Numerous research studies have been conducted to investigate the impact of modifying process parameters on enhancing the mechanical properties of components manufactured using the FDM technology. In a particular study, the tensile strength of polycarbonate parts manufactured using FDM technology was examined. The study focused on altering and investigating process variables such as air gap, raster width, and raster angle to assess their impact on the tensile strength of the parts [12].

Due to the utilization of multiple process parameters and the microstructural anisotropy resulting from the layer-by-layer building process, the mechanical properties and surface quality of final products created through FDM are comparatively inferior to those manufactured using traditional processes [13].

PLA is a thermoplastic biodegradable polymer extensively employed in various applications. Its fundamental monomer, lactic acid, is obtained through the fermentation of starch. Subsequently, lactic acid undergoes polymerization, either through gradual polycondensation or ring-opening polymerization, resulting in the formation of PLA [14]. It possesses favorable printability and strength; however, it tends to be more brittle than alternative 3D printer materials like ABS [15]. It has gained recognition as an eco-friendly alternative to traditional petroleum-based plastics due to its reduced carbon footprint and non-toxic nature [16]. PLA offers favorable characteristics for AM, including ease of use, low warping tendency, and excellent surface finish. Additionally, PLA exhibits good strength and stiffness properties, making it suitable for a wide range of applications.

Agu et al. looked at the durability of 3D printed polylactic acid (PLA) for making components that would be subject to high strain rates and impact events over the course of their design lives. They stated that the shear strength with increasing impact stress for any PLA-based 3D printed structure had been measured for the first time [17]. Currently, composite blends based on PLA are used to create 3D-printed tissue engineering scaffolds. Senatov et al. used SEM imaging to show that the structure of the PLA-based scaffolds contains interconnecting pores [18]. For the purpose of creating 3D porous, biocompatible scaffolds for diverse tissue engineering applications, PLA and a bioactive CaP glass were 3D printed together [19].

In this study, two different infill patterns were employed to fabricate tensile and compressive specimens using the extrusion-based AM process. The printed specimens were subsequently subjected to tensile and compression testing in accordance with ASTM standards to evaluate their mechanical properties.

2 Material and Methodology

PLA filament is utilized in this study to fabricate specimens for mechanical testing. Tensile specimens as per ASTM D638 Type 4 were fabricated, and for compression testing ASTM D 695 was used. Figure 1 shows the CAD model of the specimens fabricated using FDM technique along with the stl files representation.

The process of fabricating the specimens involved several steps. Firstly, the CAD model was designed using PTC Creo 7.0 software, allowing for precise customization and control over the geometry. Once the design phase was completed, the file extension was changed from .prt to .stl to make it compatible with the subsequent steps of the additive manufacturing process. In the next step, the .stl file was imported into the slicing software, OrcaSlicer, which facilitated the generation of G-code instructions. The slicing software allowed for parameter adjustments and fine-tuning of the printing settings. After the G-code extraction, it was fed into the printer to initiate

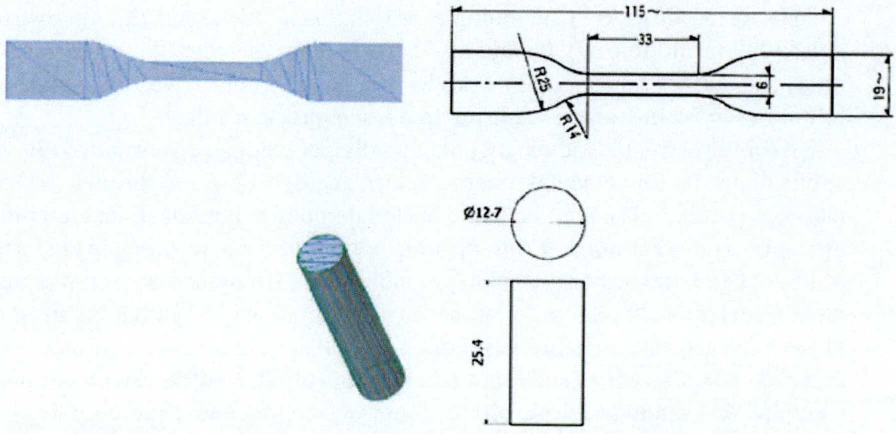

Fig. 1 Dimension of the tensile and compression specimens along with their STL representation

the printing process. Table 1 provides the necessary printing parameters, ensuring consistency and reproducibility throughout the fabrication process. To optimize the printing speed and achieve high-quality output, Klipper was utilized as the additive manufacturing firmware. Its advanced capabilities and efficient performance helped to expedite the printing process without compromising the final product's quality and integrity. For the actual printing of the specimens, an Ender3 V2 printer was employed. This specific printer model was chosen for its reliability and compatibility with the chosen printing parameters and firmware. It provided a stable and consistent platform for translating the digital design into physical specimens. The printing parameters selected for the printing the specimens are discussed in Table 1.

To explore the influence of infill patterns on the mechanical properties, two different infill patterns were employed during the fabrication process. One infill pattern was oriented at 90° raster orientation, while the other infill pattern was oriented at 0° raster orientation (Fig. 2). By utilizing these distinct infill patterns, we aimed to investigate the impact of raster orientation on the mechanical behavior of the specimens. This approach allowed us to evaluate and compare the performance of the specimens under tensile and compressive loading conditions, shedding light on the effects of infill pattern orientation on the overall mechanical properties.

Table 1 Printing parameters

Printing parameters	Value
Infill	100%
Wall count	3
Layer height	0.1 mm
Speed	150 mm/s
Nozzle temp	220 °C
Bed temp	60 °C

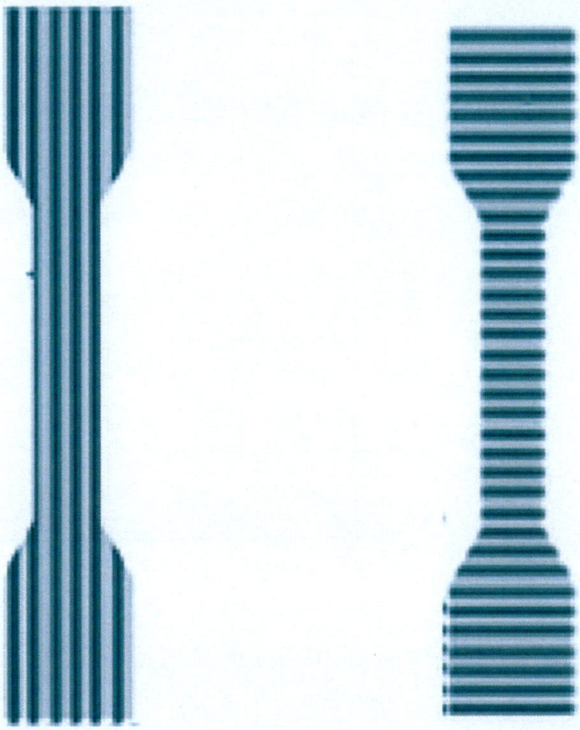

Fig. 2 Infill patterns (0° and 90°) representation

3 Results and Discussion

The tensile test was conducted on three different sets of specimens printed with two different printing patterns: 0° and 90°. The results of the tensile test, including Young's modulus and tensile strength, are summarized in Table 2.

Comparing the results of the two printing patterns, it can be observed that the specimens printed at 0° exhibited higher Young's modulus and tensile strength compared to those printed at 90°. This suggests that the 0° printed pattern has greater strength and can withstand higher loads before failure. The graph in Fig. 3 confirms this trend, showing the superior strength of the 0° printed specimens.

Furthermore, analyzing the load versus displacement graphs for each printing pattern, it can be seen that the first sample printed at 90° reached a peak load of 52 kg

Table 2 Tensile test results

Printing pattern	Young's modulus (MPa)	Tensile strength (MPa)
90°	117.306	50.391
0°	171.99	55.9475

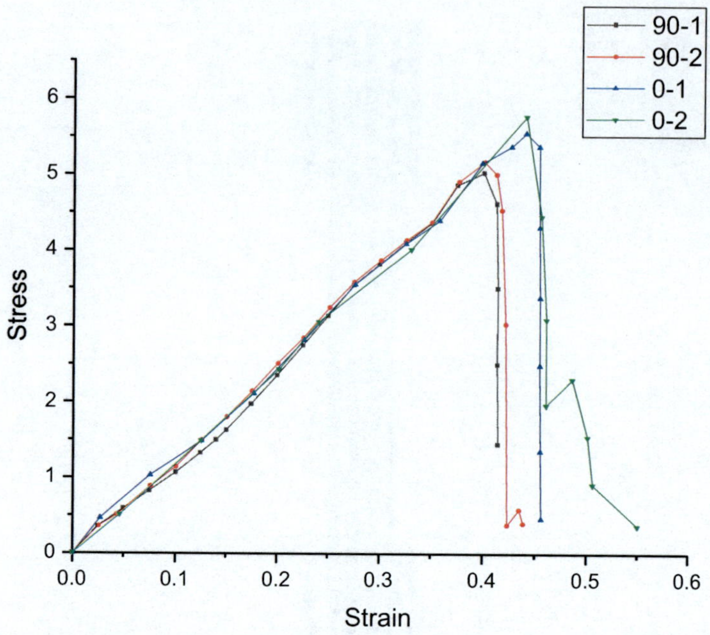

Fig. 3 Stress versus Strain graph of tensile specimens (0° and 90°)

before failure, with an elongation of 3.78 mm (Fig. 4). Similarly, the second sample printed at 0° reached a peak load of 53.2 kg with an elongation of 3.79 mm (Fig. 4).

On the other hand, the two specimens printed at 0° exhibited higher peak loads, with the first specimen reaching 57.2 kg and the second specimen reaching 59.6 kg. Both specimens had an elongation of 4.21 mm (Fig. 5). These results indicate that the 0° printed specimens have a higher load-bearing capacity and greater elongation capability compared to the 90° printed specimens.

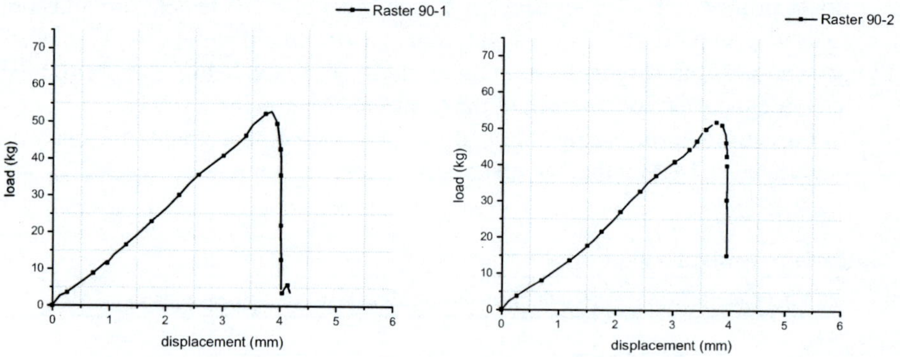

Fig. 4 Load versus displacement curve for tensile specimens (90°)

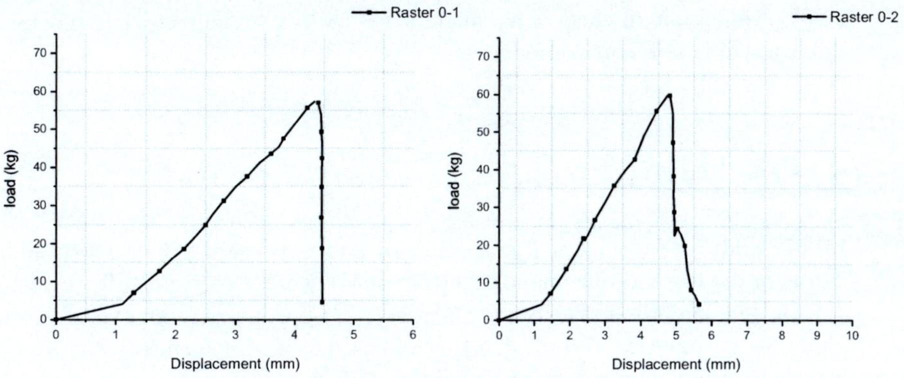

Fig. 5 Load versus displacement curve for tensile specimens (0°)

The 0° printed specimens may have better load transfer between the layers, resulting in higher tensile strength. The alignment of the layers along the loading direction allows for efficient stress transfer and load distribution, reducing the risk of localized stress concentrations and potential failure. Moreover, for the case of 90° printed specimens the bonded infill rasters may delaminate and cause the failure at much lesser stress value.

The compressive test yielded a compressive strength of the printed specimens at 22.686 MPa. The graphical representation of the compressive test results for both samples can be seen in Fig. 6. These results provide valuable insights into the mechanical properties and performance of the fabricated specimens in compression.

Overall, the tensile and compressive tests demonstrate the influence of printing pattern orientation on the mechanical behavior of the specimens. The specimens printed at a 0° raster angle exhibited higher strength and load-bearing capacity compared to those printed at 90°. These findings contribute to the understanding

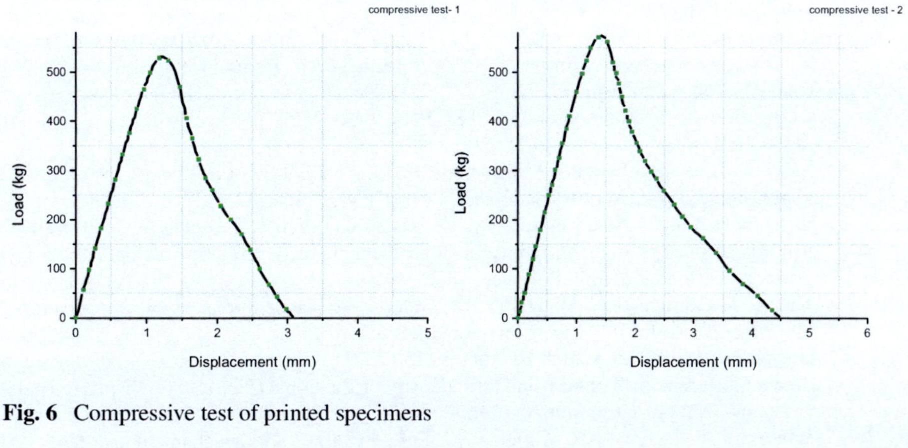

Fig. 6 Compressive test of printed specimens

and optimization of design parameters in additive manufacturing processes for biodegradable polymer components.

4 Conclusion

In this study, the mechanical properties of polylactic acid (PLA) fabricated using extrusion-based additive manufacturing (AM) were experimentally investigated. Tensile and compression tests were conducted on specimens with two different infill patterns (0° and 90°). The results showed that the specimens printed with a 0° infill pattern exhibited higher Young's modulus and tensile strength compared to the 90° pattern. The 0° printed specimens also demonstrated higher peak loads and elongation capabilities in the tensile tests. The compressive test yielded a compressive strength of 22.686 MPa for the printed specimens. Overall, the findings highlight the significant influence of infill pattern orientation on the mechanical behavior of the biodegradable polymer components. These results contribute to our understanding of the mechanical properties of AM-fabricated biodegradable polymers. Optimizing infill pattern orientation can enhance the strength and load-bearing capacity of these components. Further research and optimization efforts can drive the development of sustainable manufacturing practices and expand the application of biodegradable polymers in various industries.

References

1. Hanon MM, Marczis R, Zsidai L (2021) Influence of the AM process settings on tensile strength of PLA and HT-PLA. Periodica Polytechnica Mech Eng 65(1):38–46
2. Ferreira RTL, Amatte IC, Dutra TA, Bürger D (2017) Experimental characterization and micrography of 3D printed PLA and PLA reinforced with short carbon fibers. Compos Part B Eng 124:88–100
3. Behzadnasab M, Yousefi AA, Ebrahimibagha D, Nasiri F (2020) Effects of processing conditions on mechanical properties of PLA printed parts. Rapid Prototyping J 26(2):381–389
4. Zhao L, Jiang Z, Zhang C, Jiang Z (2021) Development model and experimental characterization of residual stress of AM PLA parts with porous structure. Appl Phys A 127:1–10
5. Aloyaydi B, Sivasankaran S, Mustafa A (2020) Investigation of infill-patterns on mechanical response of 3D printed poly-lactic-acid. Polym Test 87:106557
6. Hsueh MH, Lai CJ, Wang SH, Zeng YS, Hsieh CH, Pan CY, Huang WC (2021) Effect of printing parameters on the thermal and mechanical properties of 3d-printed pla and petg, using fused deposition modeling. Polymers 13(11):1758
7. Abbas TF, Othman FM, Ali HB (2018) Influence of layer thickness on impact property of 3D-printed PLA. Int Res J Eng Technol (Irjet) 5:1–4
8. Mishra PK, Senthil P, Adarsh S, Anoop MS (2021) An investigation to study the combined effect of different infill pattern and infill density on the impact strength of 3D printed polylactic acid parts. Compos Commun 24:100605
9. Camargo JC, Machado ÁR, Almeida EC, Silva EFMS (2019) Mechanical properties of PLA-graphene filament for FDM AM. Int J Adv Manuf Technol 103:2423–2443

10. Al Khawaja H, Alabdouli H, Alqaydi H, Mansour A, Ahmed W, Al Jassmi H (2020) Investigating the mechanical properties of 3D printed components. In: 2020 Advances in Science and Engineering Technology International Conferences (ASET). IEEE, pp 1–7
11. Khan SF, Zakaria H, Chong YL, Saad MAM, Basaruddin K (2018) Effect of infill on tensile and flexural strength of 3D printed PLA parts. In: IOP conference series: materials science and engineering, vol 429, no 1. IOP Publishing, p 012101
12. Lanzotti A, Grasso M, Staiano G, Martorelli M (2015) The impact of process parameters on mechanical properties of parts fabricated in PLA with an open-source 3-D printer. Rapid Prototyping J
13. Yang L, Li S, Li Y, Yang M, Yuan Q (2019) Experimental investigations for optimizing the extrusion parameters on FDM PLA printed parts. J Mater Eng Perform 28:169–182
14. Graupner N, Herrmann AS, Müssig J (2009) Natural and man-made cellulose fibre-reinforced poly (lactic acid)(PLA) composites: An overview about mechanical characteristics and application areas. Compos Part A Appl Sci Manuf 40(6–7):810–821
15. Naveed N (2021) Investigating the material properties and microstructural changes of fused filament fabricated PLA and tough-PLA parts. Polymers 13(9):1487
16. Brenken B, Barocio E, Favaloro A, Kunc V, Pipes RB (2018) Fused filament fabrication of fiber-reinforced polymers: a review. Add Manuf 21:1–16
17. Agu HO, Hameed A, Appleby-Thomas GJ, Wood DC (2019) The dynamic response of dense 3 dimensionally printed polylactic acid. J Dyn Behav Mater 5:377–386
18. Senatov FS, Niaza KV, Zadorozhnyy MY, Maksimkin AV, Kaloshkin SD, Estrin YZ (2016) Mechanical properties and shape memory effect of 3D-printed PLA-based porous scaffolds. J Mech Behav Biomed Mater 57:139–148
19. Serra T, Planell JA, Navarro M (2013) High-resolution PLA-based composite scaffolds via 3-D printing technology. Acta Biomater 9(3):5521–5530

Metaheuristic-Based Parametric Optimization of Abrasive Water-Jet Machining Process—A Comparative Analysis

Sunny Diyaley and Partha Protim Das

Abstract Hybrid machining (HM) processes are an effective means for increasing material removal rate, improving surface integrity, minimizing production time and tool wear. Determination of the optimal parametric settings is an important problem in hybrid machining process. The performance of these processes mainly depends on the optimal choice of its various input parameters which highly affect the responses, like material removal rate and surface roughness. In this paper, seven metaheuristic algorithms in the form of artificial bee colony (ABC), ant colony optimization (ACO), particle swarm optimization (ACO), firefly algorithm (FA), differential evolution (DE), teaching–learning-based optimization (TLBO), and elephant swarm water search algorithm (ESWSA) are employed to determine the optimal parametric settings of abrasive water-jet machining (AWJM) process, while satisfying their given sets of practical machining constraints. It is observed that ESWSA outperforms the others with respect to the derived optimal solution, consistency of the solution and convergence speed.

Keywords Optimization · Metaheuristic · Algorithm

1 Introduction

A HM process can be effectively used by combining different active energy sources acting in the processing region or by integrating process steps of two or more machining processes, else by combining different processes within one machining platform [1]. AWJM process is a HM process which integrates the mechanism of both water-jet machining (WJM) process and abrasive jet machining (AJM) process to remove material from the workpiece. In HM process, like AWJM process, the machining performance mainly depends on a set of input parameters, like mass flow

S. Diyaley (✉) · P. P. Das
Department of Mechanical Engineering, Sikkim Manipal Institute of Technology, Sikkim Manipal University, Majitar, Sikkim 737136, India
e-mail: sdiyaley@gmail.com

N. Kumar et al. (eds.), *Advances in Materials and Agile Manufacturing*, Lecture Notes in Mechanical Engineering,
https://doi.org/10.1007/978-981-99-6601-1_14

rate of the abrasives, mass flow rate of water, diameter of abrasive water-jet nozzle, feed rate of the nozzle, which directly play essential roles in achieving the desired values of its MRR. Thus, proper settings of several input parameters in AWJM process have major role to remove material rapidly with low cost and better surface finish. Conventionally, the input parameter selections in such machining processes were carried out by an experienced operator or a design engineer but they could not completely satisfy the criteria of good quality and high efficiency. In factory environment, machining tables can be considered as a better choice but cannot be used for a broad range of machining processes [2]. In most of the cases, the conventional techniques of parameter selection of machining processes have led to the partial fulfillment of the desired objectives and parameters far from optimum values. Thus, this compels the deployment of metaheuristic algorithms to explore the optimal parametric combination for hybrid machining processes. Hence, in this paper seven metaheuristics are deployed to determine the optimal parametric settings of AWJM process for yielding maximum MRR.

2 Literature Review

Recently, several literatures are reported in parametric optimization of AWJM process. Rao et al. [3] considered water-jet pressure at nozzle exit, diameter of abrasive water-jet nozzle, feed rate of nozzle, mass flow rate of water and mass flow rate of abrasives as the input parameters of an AWJM process and applied TLBO algorithm for maximization of MRR. It was observed that the TLBO algorithm outperforms the other advanced optimization techniques, like simulated annealing (SA), genetic algorithm (GA), PSO, ABC, and harmony search (HS) algorithm. Kumar et al. [4] analyzed the effect of input parameters for concurrently optimizing the three responses of an AWJM process, i.e., MRR, SR, and dimensional deviation while machining duplex stainless steel 2205 using response surface methodology (RSM). Traverse speed, stand-off distance, and abrasive flow rate were considered as the input parameters of an AWJM process and the optimal condition for maximum MRR, minimum SR value and dimensional deviation were predicted. Manoj et al. [5] considered traverse speed, water-jet pressure, and stand-off distance as the input parameters of an AWJM process while machining TiB_2 particles reinforced Al7075 composite and applied Taguchi-DEAR methodology for simultaneous optimization of three responses, i.e., MRR, SR, and taper angle. The optimal input parameters were determined, and from the experimental observation, water-jet pressure was found to be the most influential parameter. Chakraborty and Mitra [6] applied gray wolf optimizer (GWO) technique in an attempt to obtain the optimal parametric combination of an AWJM process. The results derived from applying GWO technique showed a considerable improvement in the MRR with respect to several defined constraints. It was also concluded that the optimization performance of the proposed technique was superior to the attempts made earlier for the parametric optimization of AWJM processes while applying other metaheuristic algorithms. Joel et al.

[7] successfully applied multi-objective TLBO to discover the optimal parametric settings of machining c360 brass using AWJM process by simultaneously optimizing three responses, i.e., hardness, MRR, and SR. Jagdish et al. [8] simultaneously minimized processing time and SR using hybrid spider monkey optimization (HSMO), TLBO, and GWO for an AWJM process while machining green composites. It was observed that the optimal settings by the proposed techniques were successfully verified through confirmation of experiment. Bhoi et al. [9] demonstrated the applicability of chemical reaction optimization (CRO) algorithm for searching out the optimal parametric combination of AWJM machining parameters. The proposed technique delivered better results compared to other different metaheuristics.

From the extensive review of the literatures, it can be concluded that there is an intense curiosity among researchers to implement several optimization techniques in AWJM process. They have effectively applied several optimization algorithms, like CSA, GA, PSO, ABC, HSMO, TLBO, and GWO to determine the optimal parametric combination for different AWJM process, but in most cases no research work has been conducted till date to study the comparative optimization performance of the most accepted metaheuristic algorithms in this field and to identify the best algorithm. Thus, in this paper, six well-known metaheuristics, i.e., ABC, ACO, PSO, DE, FA, TLBO, and ESWSA techniques, are adopted to determine the optimal machining parameters in AWJM. The comparative analysis reveals that the ESWSA is more robust, flexible, and reliable as compared to the other algorithms. ESWSA is easy to implement due to the smaller number of parameters adjustment required during optimization and can result in almost global optimal solution in least computational time as compared to other metaheuristics [10].

3 Metaheuristic Algorithms

Heuristic algorithms provide acceptable solutions to several optimization problems through 'trial-and-error' technique consuming a considerable amount of time for computation [11]. Moreover, they do not guarantee in obtaining global optimum and are categorized as local search methods. Hence, the concern for obtaining a global optimal solution leads the researchers toward tuning their interests into metaheuristic algorithms. A metaheuristic is an iterative method which directs a secondary heuristic by combining intelligent concepts for exploiting and exploring the search space. They are motivated by the phenomena occurring in nature and utilize algorithms which are flexible, simple, and well-accepted for fast solving of large-scale problems [12]. The term 'metaheuristic' refers to higher-level heuristic algorithms which aim to solve higher-level strategic optimization. These metaheuristic algorithms effectively search a feasible solution space which is huge enough to be entirely sampled in a reasonable time. Diversification and intensification are the two vital components of modern metaheuristics. Diversification refers to the process of exploring the search space efficiently and effectively, while intensification is the exploitation of previous solution to choose the potentially good solutions through the usage of memory or elitism or both [13]. A correct balance between these two components is very essential for any metaheuristic algorithms. Metaheuristics are primarily classified as single-solution-based and population-based algorithm. The latter has some merits over the former with respect to greater exploration capacity and lower probability of sticking at local optima. The metaheuristics, i.e., PSO, ABC, DE, FA and TLBO, and ESWSA, are some of the examples of population-based metaheuristic algorithms employed in this paper.

Optimization steps of any type of metaheuristic algorithm are provided here-in-under [14].

Step 1: In the search domain, initialize the population with random values.
Step 2: Compute the fitness for every one of the population.
Step 3: Apply evolutionary approaches, like crossover and mutation to create a new population by reproducing selected individuals.
Step 4: Go to step 2 until the criterion for termination is met.
Step 5: Step 5: End.

4 Parametric Optimization of AWJM Process

In this paper, the above-mentioned six metaheuristic algorithms are employed for parametric optimization of AWJM process, and the following values are set for their different control parameters:

ABC algorithm: Swarm size = 200, number of employed bees = 50% of the swarm size, number of onlooker bees = 50% of the swarm size, number of scouts per cycle = 1, number of cycles = 1000 and limit = 50, number of iterations = 100.
ACO algorithm: Population size = 200, sample size = 40, intensification factor = 0.5 and deviation distance ratio = 1, number of iterations = 100.
PSO algorithm: Population size = 200, inertia weight factor = 0.65, and acceleration coefficients = 1.65 and 1.75, number of iterations = 100.
FA: Number of fireflies = 200, light absorption coefficient = 1, initial randomness = 0.9, randomness factor = 0.91 and randomness reductions = 0.75, number of iterations = 100.
DE algorithm: Population size = 200, lower bound of scaling factor = 0.5, upper bound of scaling factor = 0.8, and crossover probability = 0.9, number of iterations = 100.
TLBO algorithm: Population size = 200, number of iterations = 100.
ESWSA: Population/swarm size = 200, maximum inertia weight = 0.9, minimum inertia weight = 0.2, switching probability = 0.6, number of iterations = 100.

Jain et al. [15] considered the mathematical model as developed by Hashish [16] representing the interrelationship among MRR and five input parameters of an AWJM process. The input parameters considered in that model are mass flow rate of the abrasives (M_a) (in kg/s), mass flow rate of water (M_w) (in kg/s), feed rate of the nozzle (f_n) (in mm/s), diameter of abrasive water-jet nozzle (d_{awn}) (in mm) and water-jet pressure at the nozzle exit (P_w) (in MPa). The objective function is given as:

$$\text{Maximize,} \quad \text{MMR} = d_{awn} + f_n(h_c + h_d) \tag{1}$$

where 'h_c' is the indentation depth due to cutting wear and 'h_d' is the indentation depth due to deformation wear calculated using Eqs. (2), (3), and (4).

$$h_c = \left(\frac{1.028 \times 10^{4.5}\xi}{C_k \rho_a^{0.4}}\right)\left(\frac{d_{awn}^{0.2} M_a^{0.4}}{f_n^{0.4}}\right)\left(\frac{M_w P_w^{0.5}}{M_a + M_w}\right) - \left(\frac{18.48 k_a^{2/3} \xi^{1/3}}{C_k^{1/3} f_r^{0.4}}\right)\left(\frac{M_w P^{0.5}}{M_a + M_w}\right)^{1/3}; \quad \text{if } \alpha_t \leq \alpha_0 \tag{2}$$

$$h_c = 0; \quad \text{if } \alpha_t \geq \alpha_0 \tag{3}$$

$$h_d = \frac{\eta_a d_{awn} M_a \left[K_1 M_w P_w^{0.5} - (M_a + M_w) v_{ac}\right]^2}{(1580.8\sigma_{fw}) d_{awn}^2 f_n (M_a + M_w)^2 + (K_1 C_{fw} \eta_a)\left[K_1 M_w P_w^{0.5} - (M_a + M_w) v_{ac}\right] M_a M_w P_w^{0.5}} \tag{4}$$

Angle of impingement (α_0) where the highest erosion occurs is given by

$$\alpha_0 \approx \left(\frac{0.02164 C_k^{1/3} f_r^{0.4}}{K_a^{2/3} \xi^{1/3}}\right)\left(\frac{M_a + M_w}{M_w P_w^{0.5}}\right)^{1/3} \quad \text{(degrees)} \tag{5}$$

and, angle of impingement at top of machined surface (α_t) is approximately calculated as

$$\alpha_t \approx \left(\frac{0.389 \times 10^{-4.5} \rho_a^{0.4} C_k}{\xi}\right)\left(\frac{d_{awn}^{0.8} f_n^{0.4} (M_a + M_w)}{M_a^{0.4} M_w P_w^{0.5}}\right)^{1/3} \quad \text{(degrees)} \tag{6}$$

where

$$C_k = \sqrt{3000 \sigma_{fw} f_r^{0.6} / \rho_a} \text{ (mm/s)} \tag{7}$$

and

$$K_a = 1 + m_p r_m^2 / I_p, \tag{8}$$

where m_p is mass of the abrasive particle and I_P is the moment of inertia of an abrasive particle about its center of gravity.

$$\text{Critical velocity, } v_{ac} = 5\pi^2 \frac{\sigma_{ew}^{2.5}}{\rho_a^{0.5}} \left[\frac{1 - v_a^2}{E_{Ya}} + \frac{1 - v_w^2}{E_{Yw}} \right]^2 \text{ (mm/s)} \tag{9}$$

$$K_1 = \sqrt{2} \times 10^{4.5} \xi \tag{10}$$

$$K_a = 3 \tag{11}$$

The model is subjected to a constraint for power consumption which is given by Eq. (12).

$$1.0 - \frac{P_w M_w}{P_{max}} \geq 0.0 \tag{12}$$

The variable bounds for the considered input parameters are given in Eqs. (13)–(17).

$$0.0003 \leq M_a \leq 0.08 \tag{13}$$

$$0.02 \leq M_w \leq 0.2 \tag{14}$$

$$0.2 \leq f_n \leq 25.0 \tag{15}$$

$$0.5 \leq d_{awn} \leq 5.0 \tag{16}$$

$$50.0 \leq P_w \leq 400.0 \tag{17}$$

In order to calculate the values of MRR, the values of constants are also required, as presented here-in-under [3]:

Poisson ratio of the abrasive particles (v_a) = 0.25, roundness factor of the abrasive particles (f_r) = 0.35, sphericity factor of abrasive particles (f_s) = 0.78, density of abrasive particles (ρ_a) = 3.95 × 10^{-6} kg/mm^3, Young's modulus of elasticity of abrasive particles (E_{Ya}) = 350,000 MPa, Young's modulus of elasticity of work material (E_{Yw}) = 1,14,000 MPa, proportion of abrasive grains effectively participating in machining (η_a) = 0.7, Poisson's ratio of work material (v_w) = 0.2, elastic limit of work material (σ_{ew}) = 883 MPa, flow stress of the work material (σ_{fw}) = 8142 MPa, mixing efficiency between abrasives and water (ξ) = 0.8, allowable power consumption value (P_{max}) = 56 kW, drag or skin friction coefficient for work material (C_{fw}) = 0.002.

Hasish [16] attempted GA technique to determine the optimal parameter settings resulting in maximum value of MRR as 90.257 mm^3/s. The parametric optimization of the five input parameters, i.e., M_a, M_w, f_n, d_{awn}, and P_w was reported as 0.079 kg/s, 0.141 kg/s, 23.17 mm/s, 3.726 mm, and 398.3 MPa, respectively. In this paper, seven metaheuristic algorithms are employed to determine the optimal combinations of AWJM process parameters to attain a maximum value of MRR. The results of single objective optimization of MRR are provided in Table 1, along with the comparison of the optimal solutions with those as achieved by the other algorithms. It is observed that ESWSA outperforms over the others with respect to derived optimal solutions and consistency of the solution with respect to standard deviation (SD) value. The application of ESWSA algorithm also showed an increase in MRR to 317.780 mm^3/s. Thus, for achieving maximum MRR, a mix of $M_a = 0.0793$ kg/s, $M_w = 0.14$ kg/s, $f_n = 22.189$ mm/s, $d_{awn} = 3.886$ mm, and $P_w = 397.4$ MPa can be set. While solving this optimization problem of maximization of MRR for AWJM process using seven metaheuristic algorithms, the constraint for power consumption given in Eq. (12) was not violated by any of the algorithms as exhibited in Table 2. The superiority of ESWSA algorithm can also be well validated from the convergence diagrams for MRR, as shown in Fig. 1a. In this figure, it can be clearly seen that ESWSA algorithm quickly converges to almost global optimal solutions as compared to the remaining algorithms. The uniqueness of TLBO algorithm over the other considered algorithms with respect to variability of the derived solution is also validated by presenting box plot of the derived solution, as presented in Fig. 1b. These box plots are the statistical methods to portray statistical data graphically where a rectangle is drawn to signify the inter-quartile range of the derived solutions. The median value is shown by a small box inside the rectangle and two lines (whiskers) beyond each box extend to the least and maximum value of the observation.

In Fig. 2, scatter plots are presented in order to depict the variations of MRR with respect to various input parameters of the AWJM process. The optimality of the solution mentioned above can be confirmed with the help of these plots. From these scatter plots, it can be revealed that MRR increases as M_a, M_w, f_n, d_{awn}, and P_w increase as already reported by Hashish [16]. It was assumed that no material removal occurs in AWJM process, if the critical impact angle (α_o) is lower than the angle of impingement at the top of the surface (α_t) and only deformation wear is responsible for material removal. Thus, it was also reported by Hashish [16] that in higher values of feed rate of the nozzle and diameter of abrasive water-jet nozzle, the MRR values suddenly drop as the angle of impingement at the top surface goes beyond the critical impact angle.

5 Conclusions

In this paper, an attempt is made to determine the optimal parametric combinations using seven metaheuristic algorithms for AWJM process, so as to attain their enhanced machining performance. The optimization performance of the considered algorithms is compared among each other. It is found that ESWSA determines the maximum MRR for AWJM process while satisfying their practical machining constraints. As compared to the remaining algorithms, convergence rate of ESWSA is fast toward achieving global optimal solution with minimum variability and high

Table 1 Results of single objective optimization for MRR

Algorithm	Mean	SD	M_a (kg/s)	M_w (kg/s)	f_n (mm/s)	d_{awn} (mm)	P_w (MPa)	α_o	α_t	h_c	h_d	MRR (mm^3/s)
ABC	307.26	0.337	0.0777	0.14	21.145	3.899	397.4	0.324	0.288	5.213	3.014	307.309
ACO	306.08	0.341	0.0777	0.14	21.150	3.899	398.7	0.321	0.301	5.100	2.997	306.165
PSO	313.08	0.607	0.0769	0.14	23.150	3.799	399.3	0.346	0.298	5.145	3.189	313.251
FA	309.62	0.802	0.0770	0.13	22.711	3.884	399.4	0.329	0.289	5.128	3.147	309.919
DE	315.14	0.774	0.0780	0.14	22.144	3.884	399.4	0.356	0.314	5.257	3.214	315.310
TLBO	316.28	0.213	0.0780	0.13	22.150	3.899	398.2	0.388	0.345	5.722	3.325	316.326
ESWSA	317.72	0.223	0.0793	0.14	22.189	3.886	397.4	0.394	0.358	5.875	3.358	317.780

Table 2 Values of constraint for power consumption

Algorithm	Constraint
	≥ 0
ABC	0.006
ACO	0.074
PSO	0.013
FA	0.007
DE	0.001
TLBO	0.004
ESWA	0.015

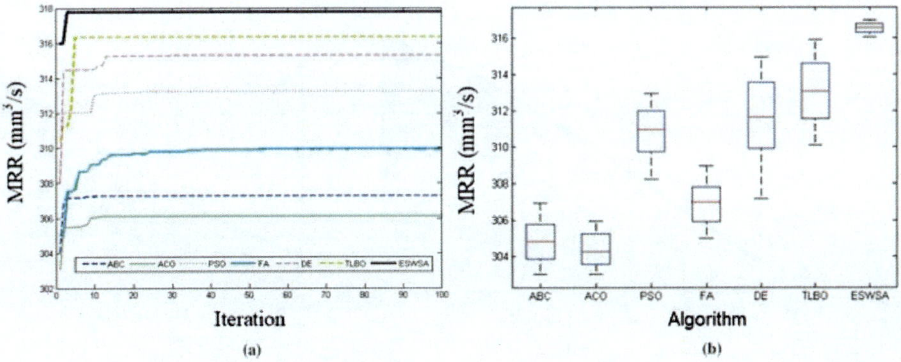

Fig. 1 **a** Convergence diagram for AWJM process. **b** Box plot for the considered optimization techniques

consistency of the derived solutions. Hence, it can be concluded that this algorithm can be successfully applied for determining the optimal parametric combination of other hybrid machining processes, like electrochemical discharge machining (ECDM), electrochemical honing (ECH), and electrochemical grinding (ECG). The process engineers can now choose the best mix of AWJM input parameters without depending on machining data handbook. The application of ESWSA for optimization other hybrid machining processes may be another scope for further study.

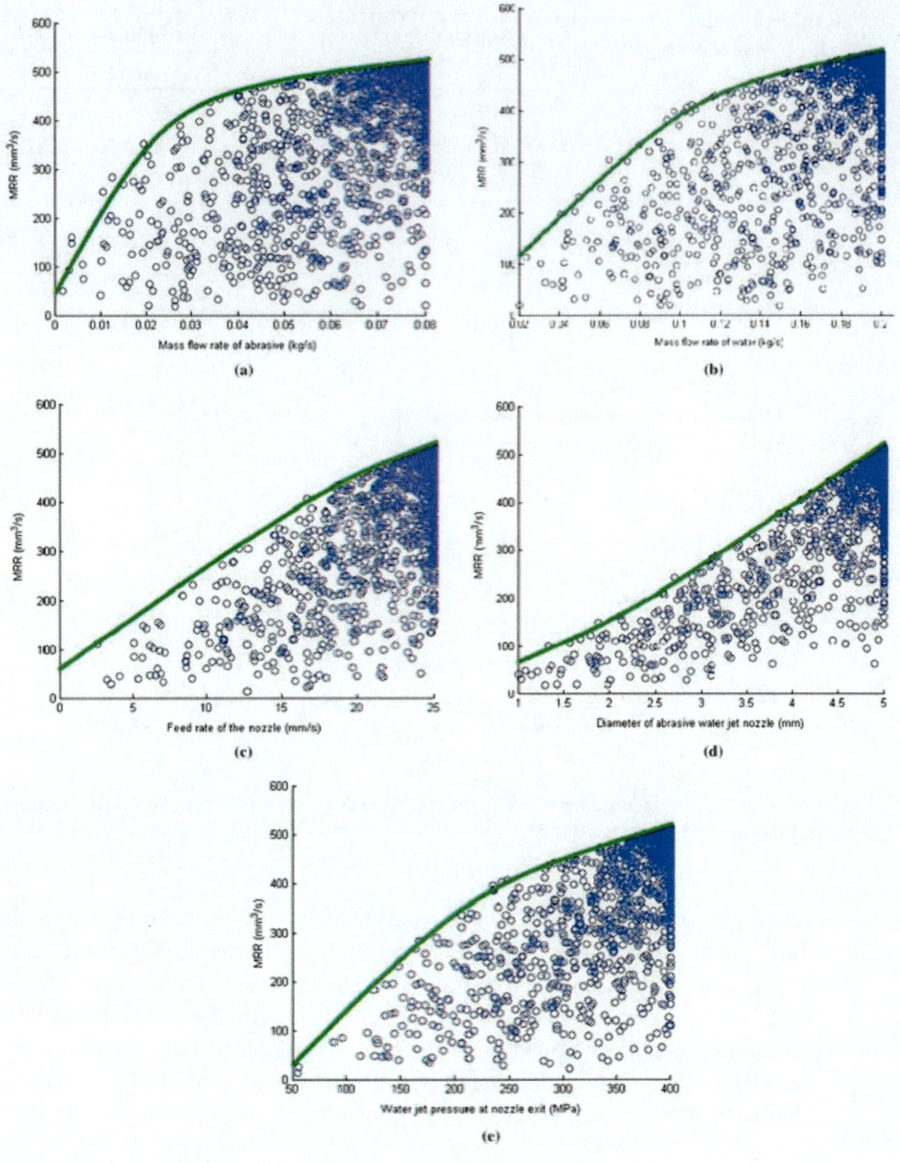

Fig. 2 Effects of various input parameters on MRR

References

1. Saxena KK, Bellotti M, Qian J, Reynaerts D, Lauwers B, Luo X (2018) Overview of hybrid machining processes: hybrid machining-theory, methods, and case studies. Elsevier, USA
2. Jain NK, Jain VK, Deb K (2007) Optimization of process parameters of mechanical type advanced machining processes using genetic algorithms. Int J Mach Tools Manuf 47(6):900–919

3. Pawar PJ, Rao RV (2013) Parameter optimization of machining processes using teaching-learning based optimization algorithm. Int J Adv Manuf Technol **67**(5–8):995–1006
4. Kumar GK, Arunachalam M, Abinash B, Kumar BK (2018) Optimization of abrasive water jet machining process parameters for duplex stainless steel-2205 by using response surface methodology. Int J Sci Res Mech Mater Eng 2(3):17–28
5. Manoj M, Jinu GR, Muthuramalingam T (2018) Multi response optimization of AWJM process parameters on machining TiB_2 particles reinforced Al7075 composite using Taguchi-DEAR methodology. SILICON 10(5):2287–2293
6. Chakraborty S, Mitra A (2018) Parametric optimization of abrasive water-jet machining processes using grey wolf optimizer. Mater Manuf Processes 33(13):1471–1482
7. Joel C, Joel L, Muthukumaran S, Shanthini PM (2021) Parametric optimization of abrasive water jet machining of C360 brass using MOTLBO. Mater Today Proc 37:1905–1910
8. Jagadish, Patel GCM, Sibalija TV (2022) Abrasive water jet machining for a high-quality green composite: the soft computing strategy for modeling and optimization. J Braz Soc Mech Sci Eng 44:83
9. Bhoi NK, Singh H, Pratap S, Jain PK (2022) Chemical reaction optimization algorithm for machining parameter of abrasive water jet cutting. Opsearch 59(1):350–363
10. Mandal S (2018) Elephant swarm water search algorithm for global optimization. Sādhanā 43(1):1–21
11. Khosravanian R, Mansouri V, Wood DA, Alipour MR (2018) A comparative study of several metaheuristic algorithms for optimizing complex 3-D well-path designs. J Pet Explor Prod Technol 8(4):1487–1503
12. Diyaley S, Chakraborty S (2019) Optimization of multi-pass face milling parameters using metaheuristic algorithms. Facta Universitatis Ser Mech Eng 17(3):365–383
13. Yang XS (2009) Harmony search as a metaheuristic algorithm. Studies in Computational Intelligence, Springer, Berlin
14. Madić M, Marković D, Radovanović M (2013) Comparison of meta-heuristic algorithms for solving machining optimization problems. Facta Universitatis Ser Mech Eng 11(1):29–44
15. Jain NK, Jain VK, Jha S (2007) Parametric optimization of advanced fine-finishing processes. Int J Adv Manuf Technol 34(11–12):1191–1213
16. Hashish M (1989) A model for abrasive water jet (AWJ) machining. J Eng Mater Technol 111(2):154–162

Vibro-Acoustics Analysis for Tool Wear Monitoring During the Turning Operation

Saurabh Tiwari, Ashish Kumar, and M. Amarnath

Abstract Detecting and monitoring of wear on cutting tools is essential in metal-cutting processes as excessive wear led to deteriorate machined surfaces and dimensional distortions in the manufactured components. To address this issue, various research endeavors have focused on the development of tool condition monitoring systems. These systems aim to accurately identify and track the wear of cutting tools, enabling timely interventions and maintenance to minimize production defects and associated expenses. This paper presents an effective tool wear monitoring technique by utilizing vibro-acoustic signature analysis. It employs the cutting tool vibration signals and the audible sound emitted during the machining operation. Based on the analysis of vibro-acoustic signals conducted in both the time and frequency domains, it was observed that certain components of the measured signals exhibited a strong correlation with the accumulated wear on the cutting tool. Furthermore, this technique may also be utilized to predict the tool wear at various cutting environments.

Keywords Tool wear monitoring · Vibro-acoustic analysis · Fast Fourier transform

1 Introduction

During machining, the shear contact between the cutting tool, workpiece, and chip extraction due to pressure leads to changes in the tool's shape. This process occurs continuously resulting in tool wear or sometimes abruptly causing tool breakage [1].

S. Tiwari (✉) · A. Kumar · M. Amarnath
Department of Mechanical Engineering, Indian Institute of Information Technology, Design, and Manufacturing, Jabalpur, Madhya Pradesh 482005, India
e-mail: saurabhgtiwari@gmail.com

A. Kumar
e-mail: 1913603@iiitdmj.ac.in

M. Amarnath
e-mail: amarnath@iiitdmj.ac.in

N. Kumar et al. (eds.), *Advances in Materials and Agile Manufacturing*,
Lecture Notes in Mechanical Engineering,
https://doi.org/10.1007/978-981-99-6601-1_15

Excessive wear on the cutting tool results in significant degradation of surface finish quality, and dimensional accuracy which may lead to the rejection of manufactured components. Tool wear has a direct impact on the cost of the finished product. In the worst-case scenario, it could even result in catastrophic tool failure which can potentially cause damage to the machine tool and incur high replacement costs [2]. Therefore, maintaining vigilant supervision of tool wear is essential to avoid these undesirable outcomes. It is crucial to closely monitor the levels of tool wear to prevent any negative consequences. Tool wear monitoring (TWM) during the metal-cutting operation can be a substantial prospect for effective machining. TWM can be utilized to generate economic tool replacement policies, adaptive control, and efficient manufacturing processes. Modern advanced machining techniques include TWM for machining performance improvement [3]. However, a single sensor may not provide sufficient sensory information about a tool's wear condition. Therefore, the identification of tool wear can be estimated more accurately with the fusion of information from multiple sensors and correlating the signal properties with the tool wear.

On the other hand, during the metal-cutting operation, the workpiece–tool-chip interface encounters a higher cutting temperature due to the friction between the contact surfaces. This has a direct impact on surface roughness, dimensional accuracy, and tool life [4]. To dissipate this high frictional temperature, various cooling techniques have been developed. Mineral oil-based cutting fluids (MO) are commonly used as a coolant/lubricant to carry away the heat generated at the workpiece-tool-chip interaction zone. This traditional approach of using mineral oil-based cutting fluids resolves the metal-cutting issues to some extent; however, continuous use of these cutting fluids led to harmful effects on human health and the environmental ecology. The volatilization and direct contact of these cutting fluids may cause respiratory and skin-related health issues such as chronic bronchitis, asthma, airway irritation, and folliculitis. On the other hand, MO also possesses low biodegradability characteristics as a result oil residues persist for a longer duration in water and soil. The additives blended with MO might result in severe pollution. Hence, the application of MO-based cutting fluids results in a slew of environmental, financial, and health difficulties. Hence, as an alternative to MO-based cutting fluid, the researcher has demonstrated the sustainability of vegetable oils as a coolant/lubricant which has no negative consequences on the environment and human health. Cutting fluids prepared by using vegetable oil offers numerous advantages. These fluids are biodegradable, less hazardous to the environment, and therefore sustainable as they are extracted from agricultural products [5–7].

This paper presented an experimental approach for developing a cost-effective TWM system based on vibro-acoustic signature analysis. The turning experiment was carried out on EN8 steel with carbide tool insert at three different cutting conditions, viz. dry, mineral-based oil (MO), and vegetable oil as a cutting fluid. Palm oil (PO) is used as a coolant/lubricant for vegetable oil-based cutting condition. The oxidation affinity of vegetable oils depends on the concentration of oleic acid. The lower the percentage of oleic acid, the lowest will be the oxygen affinity of vegetable oil. The castor oil will have the lowest oleic acid percentage; however, its higher viscosity shows poor flowability. Palm oil, which includes 38–41% oleic acid, shows a lower propensity for auto-oxidation than other vegetable oils. The balanced amount of polyunsaturated molecules present in vegetable oil provides high oxidation stability [8]. Therefore, PO has been considered as a coolant/lubricant in this study to obtain desired machining performance. The cutting tool vibration of the tool holder in the tangential direction of the workpiece rotation and the near-field acoustic signals were recorded using the accelerometer and microphone.

2 Tool Condition Monitoring in Turning

Tool condition monitoring (TCM) provides the tool wear conditions during the machining process with minimal human intervention. The worn tool can be replaced in time by utilizing an effective TCM technique which results in the reduction of cost and waste. There are various techniques available for TCM such as cutting force measurement, temperature measurement, acoustic emission, and vibration monitoring. Among the various techniques, vibration acquired by the accelerometer showed a significant correlation with the tool wear. The increase in tool wear resulted in high-frequency energy. On the other hand, variations in cutting parameters and cutting conditions affect the noise during the metal-cutting operation [9, 10].

2.1 Vibration Analysis

Vibration is caused by periodic wave oscillations which result in cyclic fluctuations in the dynamic components of the cutting forces [11]. The simplest mathematical model considering a single degree of freedom (SDoF) orthogonal turning process with a flexible tool and relatively rigid workpiece is shown in Fig. 1.

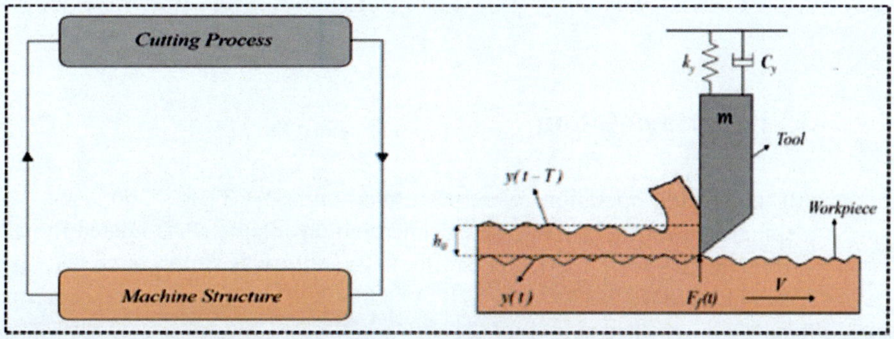

Fig. 1 Relative motion between the tool and the workpiece in turning

The tool parameters are the mass (m), stiffness (k), and damping coefficient (c), respectively. Here, V is the cutting velocity of the workpiece, $y(t)$ is the wave generated during the current revolution, and $y(t - T)$ is the wave generated during the previous revolution of the workpiece. For an SDoF dynamic system, the equation of motion in the radial direction can be written as [12]:

$$m\ddot{x}(t) + c\dot{x}(t) + kx(t) = F_f(t) \tag{1}$$

$$F_f(t) = K_f \times h_0 \times [y(t - T) - y(t)]. \tag{2}$$

Here, $[y(t - T) - y(t)]$ is the dynamic chip thickness due to tool vibration, T is the time delay between the current time and the previous time, K_f is the cutting coefficient in the feed direction, and h_0 is the chip width (width of cut). The cutting tool vibration signals acquired by the sensors are in the time waveform. Further, Fourier transform (FT) is used to convert time waveform to the frequency domain in which tool wear frequency has appeared on the amplitude of the natural frequency of the tool holder. These frequency domain vibration signals were sensitive to tool wear [13]. Fourier analysis decomposes a signal into its frequency components to determine its relative signal strengths. The Fourier transform of a time-domain signal is given as:

$$F(\omega) = \int_{-\infty}^{\infty} x(t)e^{-j\omega t}\,\mathrm{d}t. \tag{3}$$

From the transform $F(\omega)$, the time-domain signal can be obtained by inverse FFT as

$$x(t) = \frac{1}{2\pi}\int_{-\infty}^{\infty} F(\omega)e^{-j\omega t}\,\mathrm{d}\omega. \tag{4}$$

3 Experimentation

All the turning operations were performed on a PACTURN 360 center lathe with a spindle power of 20 kW. The commercially available PVD-coated rhombic ZCC carbide inserts were used as a cutting tool, and it was rigidly mounted on an ISO-designated (SDJCR 2020 K11 WIDAX) tool holder. The complete experimentation details were presented in Table 1.

The general purpose EN8 steel was used as a workpiece material having dimensions of 400 mm in length and 60 mm in diameter. All the cutting parameters such as

Table 1 Technical specification of work material, cutting parameters, and cooling conditions

Category	Specifications
Machine tool	PACTURN-360 center lathe (2.4 kW, spindle speed: 45–1800 RPM)
Material	EN8 steel (C = 0.36–0.45%, Mn = 0.6–1%, Si = 0.2–0.3%, S = 0.025%, P = 0.015%)
Dimension	Ø 60 × 400 mm
Cutting tool insert	Rhombic ZCC carbide insert DCMT11T308 (PVD-TiAlN/TlN)
Feed rate, f (mm/rev)	0.10
Cutting speed, v_c (m/min)	125
Depth of cut, a_p (mm)	1
Machining time	60 min
Cutting conditions	Dry, MO, PO
Coolant flow rate	10 ml/min (gravity feed)

cutting speed, feed rate, and depth of cut were kept constant and selected based on the tool manufacturer's recommendation. The turning tests were conducted in three different cutting conditions, viz. dry, MO, and PO as a cutting fluid. Each experiment employed a new carbide tool insert. A schematic representation of the experimental setup is depicted in Fig. 2.

A piezoelectric uniaxial accelerometer sensor was magnet mounted on the top surface of the tool holder to measure the cutting tool vibration during the turning operation. A near-field ICP microphone was placed closer to the workpiece tool interaction zone to acquire the sound during the turning. The vibration and sound signals were sampled at 10 and 20 kHz of sampling frequencies. Then, each of the signals was filtered and analyzed on the computer. The tool wear was measured

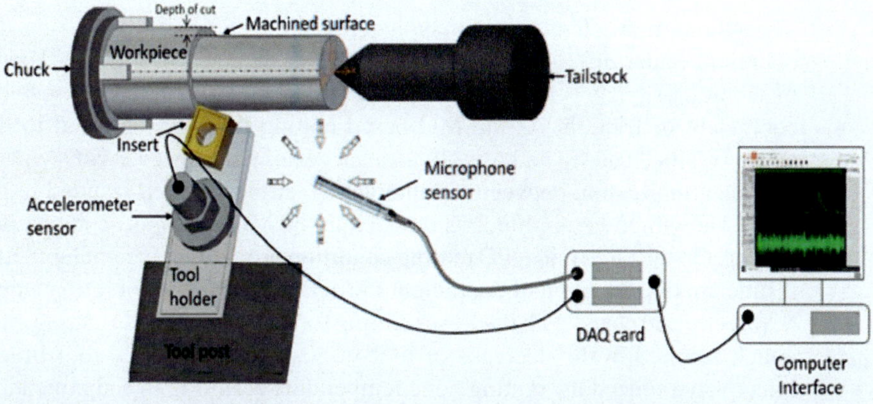

Fig. 2 Schematic view of the experimental setup

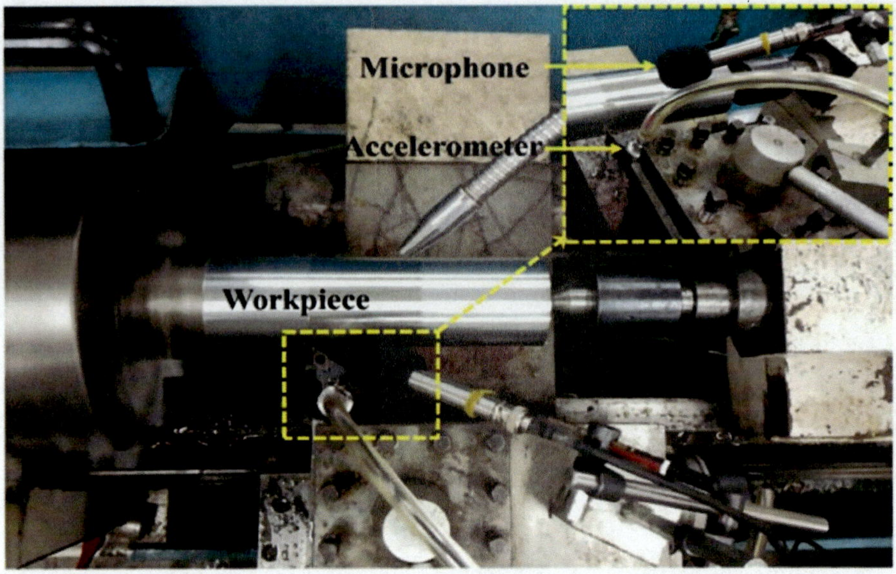

Fig. 3 Pictorial representation of the experimental setup

with the help of a USB digital microscope after 60 min of cutting operation. The photographic representation of an experimental setup is depicted in Fig. 3.

4 Results and Discussions

4.1 Tool Vibration Analysis

The vibration acceleration signals acquired in time waveform during the turning operation under different cutting conditions is depicted in Fig. 4a. The amplitude of vibration signals is highest during the dry-turning operation because of the high coefficient of friction. As the MO-based cutting fluid was allowed to flow in the cutting region, it dissipated some of the heat generated. However, MO will not provide sufficient lubrication between the interacting surfaces which resulted in higher tool wear. The vibration amplitude is reduced in the MO cutting as compared to the dry cutting. On the other hand, PO cutting condition provides sufficient lubrication which resulted in the reduction of coefficient of friction. The lesser vibration amplitude can be directly correlated with the lesser cutting forces during the PO cutting environment which resulted in the decrease of tool wear and an increased tool life. Also, PO effectively reduced the cutting zone temperature resulting in a significant reduction of vibration amplitude.

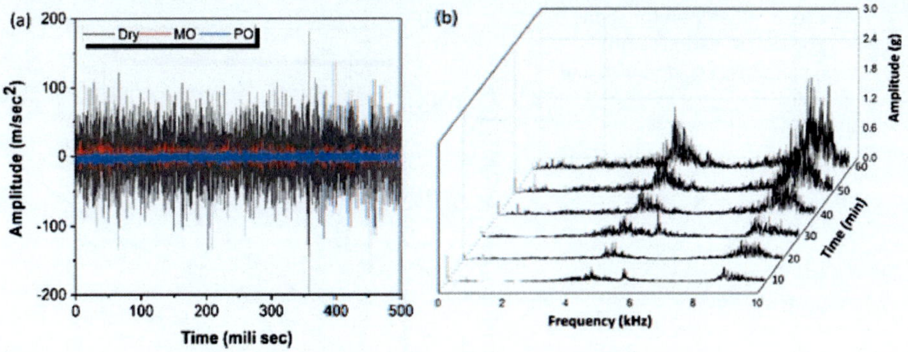

Fig. 4 **a** Time-domain plot for different cutting conditions, **b** Frequency spectrum of vibration signals for dry cutting with machining time

The direct signals measured by the vibration sensor are in the time domain which is first converted into the frequency domain with the help of fast Fourier transform (FFT). The frequency spectrum of vibratory signals under dry-cutting conditions is depicted in Fig. 4b. The spectrum showed the main peak frequency bands at 3–5 kHz which indicates the fundamental resonant frequency of the tool holder. The frequency amplitudes were increased with the increase in flank wear on the cutting tool inserts. This shows that the frequency spectrum can be directly correlated with tool wear.

4.2 Sound Analysis

Each cutting process generated a noise depending on the vibration of the machine tool and the workpiece [14]. The graph between the RMS values of sound pressure signals with the machining time is depicted in Fig. 5. It has been observed that the RMS values of the sound pressure signals are increased with the increase in tool wear. The measured acoustics signals are more sensitive to the tool wear and can be used to predict the tool condition during the turning operation. The tool wear condition can be effectively monitored and correlated with the RMS values of the sound pressure signals. This proves that it can be a very cost-effective and non-contact sensor technique for TCM during the turning operation.

4.3 Tool Wear Analysis

The life of cutting tools is considered as an essential economic aspect because of their importance in enhancing production. The flank wear developed on the tool flank surface affects the machining performance. The rubbing of a cutting tool on the workpiece causes flank wear on the clearance surface of the tool inserts, the

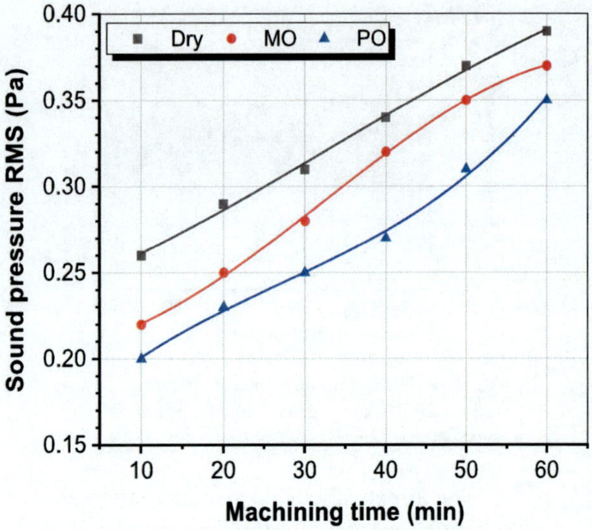

Fig. 5 RMS plot for sound pressure at various cutting conditions, viz. dry, wet, and palm oil

abrasion wear being the major wear mechanism on the flank surface [15]. Figure 6 depicts the progression of flank wear with the machining time under different cutting conditions. Under a PO-cutting environment, flank wear is minimum when compared to dry and MO conditions. The application of palm oil as a lubricant at the tool–workpiece interface resulted in a significant reduction of cutting forces and effective temperature management at the cutting zone which extends the tool life. The unique properties of PO such as excellent lubricity, oxidation stability, thermal stability, and high viscosity index result in a significant reduction in adhesion, abrasion, and diffusion wear processes thereby causing a reduction in tool wear as compared to the dry and MO cutting condition.

5 Conclusions

The analysis and correlation of vibro-acoustic signal features to cutting tool wear have been carried out. The temporal waveform of the measured signals was found to be more sensitive to the cutting condition, while frequency domain features were correlated with the tool wear. The increase in tool wear showed that the cutting conditions are influenced by the vibration and noise generated during the turning operation. The effect of tool wear was prominent from the frequency spectrum of the vibration signals. In general, the RMS value of the sound pressure amplitude increased with the increase in tool flank wear. The vibro-acoustic TCM technique is more favorable for different cutting conditions.

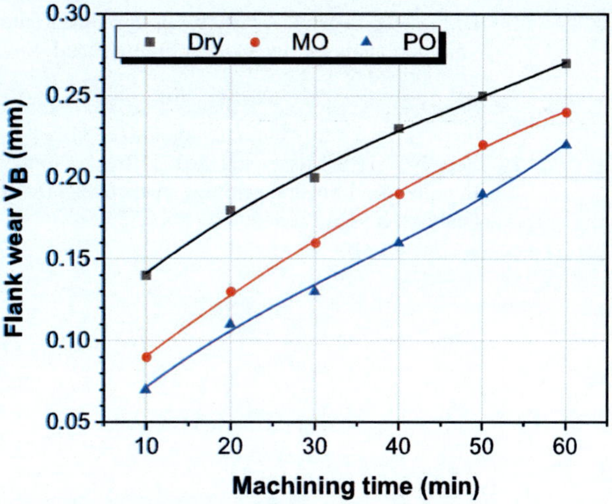

Fig. 6 Progression of tool wear with machining time

References

1. Zhu K, Wong YS, Hong GS (2009) Wavelet analysis of sensor signals for tool condition monitoring: a review and some new results. Int J Mach Tools Manuf 49:537–553. https://doi.org/10.1016/j.ijmachtools.2009.02.003
2. Dimla Snr DE (2002) The correlation of vibration signal features to cutting tool wear in a metal turning operation. Int J Adv Manuf Technol 19:705–713. https://doi.org/10.1007/s001700200080
3. Siddhpura M, Paurobally R (2013) Experimental investigation of chatter vibrations in facing and turning processes. Int J Mech Aerosp Ind Mechatron Manuf Eng 7:84–89
4. Bhuiyan MSH, Choudhury IA, Dahari M (2014) Monitoring the tool wear, surface roughness and chip formation occurrences using multiple sensors in turning. J Manuf Syst 33:476–487. https://doi.org/10.1016/j.jmsy.2014.04.005
5. Wang X, Li C, Zhang Y, Ding W, Yang M, Gao T et al (2020) Vegetable oil-based nanofluid minimum quantity lubrication turning: academic review and perspectives. J Manuf Process 59:76–97
6. Hamran NN, Ghani JA, Ramli R, Haron CC (2020) A review on recent development of minimum quantity lubrication for sustainable machining. J Clean Prod 268:122165
7. Sen B, Mia M, Krolczyk GM, Mandal UK, Mondal SP (2021) Eco-friendly cutting fluids in minimum quantity lubrication assisted machining: a review on the perception of sustainable manufacturing. Int J Precis Eng Manuf Green Technol 8:249–280
8. Sen B, Gupta MK, Mia M, Mandal UK, Mondal SP (2020) Wear behaviour of TiAlN coated solid carbide end-mill under alumina enriched minimum quantity palm oil-based lubricating condition. Tribol Int 148:106310
9. Chelladurai H, Jain VK, Vyas NS (2008) Development of a cutting tool condition monitoring system for high speed turning operation by vibration and strain analysis. Int J Adv Manuf Technol 37:471–485
10. Kopač J, Šali S (2001) Tool wear monitoring during the turning process. J Mater Process Technol 113:312–316
11. Snr DD (2001) Correlation of cutting force features with tool wear in a metal turning operation. Proc Inst Mech Eng Part B J Eng Manuf 215:435–440
12. Prasad BS, Babu MP, Reddy YR (2014) Evaluation of correlation between vibration signal features and three-dimensional finite element simulations to predict cutting tool wear in turning operation 2014. https://doi.org/10.1177/0954405414554018

13. Dimla DE, Lister PM (2000) On-line metal cutting tool condition monitoring: I: force and vibration analyses. Int J Mach Tools Manuf 40:739–768. https://doi.org/10.1016/S0890-6955(99)00084-X
14. Seemuang N, McLeay T, Slatter T (2016) Using spindle noise to monitor tool wear in a turning process. Int J Adv Manuf Technol 86:2781–2790
15. Fang N, Pai PS, Mosquea S (2011) Effect of tool edge wear on the cutting forces and vibrations in high-speed finish machining of Inconel 718: an experimental study and wavelet transform analysis. Int J Adv Manuf Technol 52:65–77

Advanced Plasma Polishing Process: Principle, Recent Applications, Challenges, and Future Scope

Hari Narayan Singh Yadav and Manas Das

Abstract In current years, optical components are the most important and necessary components of making electronic chips, semiconductors, lenses, and mirrors. In those fields, the quality of the surface is very important. Micro-cracks, lattice disturbances, and other flaws are always present on the final surfaces of components manufactured in traditional contact machining methods. In order to satisfy all those drawbacks, an advanced plasma polishing process, named medium-pressure plasma polishing (MPPP) process, has been developed. It removes surface and subsurface damage while polishing of optical components. The present article focused on a comprehensive literature review of the proposed process, with principles and applications. Moreover, this article addresses the investigation of surface roughness, morphology, and chemical composition before and after the MPPP process using 3D optical profiler, FESEM, and EDX analysis, respectively. The surface roughness of fused silica has been slightly increased from initial surface roughness (R_a) 0.35 to final R_a 0.37 μm after plasma processing. Further, the result shows the morphology has been improved after processing. EDX result shows the presence of fluorine on the surface of fused silica, and it shows the reaction has been involved during processing. Moreover, the research challenges and future scope have also been discussed.

Keywords Plasma polishing · Material removal · Surface roughness · FESEM · EDX · Optical material

Nomenclature

MPPP	Medium-pressure plasma process
PAP	Plasma-assisted polishing
APPP	Atmospheric pressure plasma process

H. N. S. Yadav (✉) · M. Das
Department of Mechanical Engineering, Indian Institute of Technology Guwahati, Guwahati, Assam 781039, India
e-mail: h.narayan@iitg.ac.in

N. Kumar et al. (eds.), *Advances in Materials and Agile Manufacturing*, Lecture Notes in Mechanical Engineering,
https://doi.org/10.1007/978-981-99-6601-1_16

SF_6	Sulfur hexafluoride
RF	Radio-frequency
MFM	Mass flow meters
EDX	Energy-dispersive X-ray
FESEM	Field-emission scanning electron microscopy
MRR	Material removal rate
sccm	Standard cubic centimeter per minute
APPJ	Atmospheric pressure plasma jet
AFM	Atomic force microscopy
XPS	X-ray photoelectron spectroscopy

1 Introduction

In recent years, the demand for optical components has been increasing due to the development of electronics-related manufacturing industries. Also, the usage of optical components is increasing due to the rapid growth of science and the evolution of modern technologies. So, considering these factors, the surface finishing of optical components must be in the range of nanofinish and ultra-smooth surfaces [1]. The optical components play an important role in producing electronic chips and semiconductor-correlated products. The commonly used optical components like lenses, telecommunication lasers, and mirrors [2]. Due to its properties, the optical components are majorly used in aeronautical navigation, aerostatic and electronic components. The main reason for using optical components in those fields is because of their individual physical properties [1]. Numerous optical materials are used in various applications like precision lenses, synchrotrons, optics windows, wave plates, optical domes, infrared optics, and precision mirrors. In addition to optical fields, fused silica is commonly used in laser-related components, aeronautical components, electronics chips, etc. These optical materials' outstanding finishing quality is crucial since these parts are utilized in extremely precise equipment such as space telescopes, metrology, machine inspection, surveillance, projection, and thermal imaging, among other things. The components of optical are categorized into many categories; materials mentioned below are very much used in the fields of aeronautical, surveillance metrology, electronics, manufacturing section, etc. The subsequent classifications are the substance which is crystalline, like calcium fluoride, and glasses, like fused silica. The substance has a low thermal expansion, like zerodur, and metals like copper and aluminum. The reason behind using fused silica is because of its unique physical properties. The properties, like more resistance and withstanding high temperature [3], will not be chemically reactive and also transparent to visible light. Nanofinishing through unconventional techniques has emerged as a novel technology for optical material processing [4]. Material surface quality is critical for science and industry. In addition to this, high processing efficiency is necessary to process the components. The atmospheric pressure plasma process (APPP) technique can easily polish complex interior shapes, freeform, and targeted areas. The potential for using these sophisticated finishing procedures to polish various complicated freeform components constructed of diverse materials is examined.

The surface and substrate properties significantly impact the life and efficiency of the optical device [5]. The final surfaces of components created by the contact machining process are constantly subject to flaws such as micro-cracks, pits, adhered material, and lattice disturbances. In order to satisfy and overcome all those drawbacks, researchers put more effect into developing advanced technology to make ultra-smooth surface finishing. A non-contact medium-pressure plasma process (MPPP) has been developed [6]. It is one of the very important methods for surface finishing of optical components, i.e., lenses, mirrors, prism, etc. [7]. Using chemical and physical actions, it can remove the material atom by atom level. This newly developing technique gives us a very good surface finish without any surface contamination and zero sub-surface defect on optical components [8]. Arnold et al. [9] proposed a plasma jet-based process, and it could be applied for surface figures or shaping error modification on an optical substrate. Moreover, it also discussed outlooks of the surface with plasma interaction. Paetzelt et al. [10] reported plasma jet polishing applies to the highly localized working area, i.e., plane, spherical, aspheric, and micro-structured substrates. Jin et al. [11] reported plasma finishing processes on a fused silica substrate. Also, an investigation is conducted into the surface characterizations of fused silica using Ar/CF_4 plasma at atmospheric pressure. Moreover, atomic force microscopy (AFM) and X-ray photoelectron spectroscopy (XPS) are employed to examine the morphology and chemical development of fused silica surfaces. Gerhard et al. [12] reported the polishing of fused silica and investigated the surface and near-surface effects at atmospheric pressure. The plasma process significantly reduces the transmission of the investigated glass samples. Additionally, ellipsometry shows a reduction in the superficial index of refraction of around 3.66% at a wavelength of 636.7 nm. A reduction in surface polarity of 30.23% is found using surface energy measurements. Cui et al. [13] reported a combined theoretical and experimental investigation into the influence of packing surface structure and plasma discharge on SF_6 degradation in γ-Al_2O_3 packing DBD system. Experiment results reveal that the hydration impact of the surface (due to moisture) and the presence of reactive gases in the plasma can considerably reduce SF_6 degradation, but they have little influence on discharge behavior. Graves [14] observed that non-equilibrium, low-pressure, weakly to partly ionized gas discharge plasmas are employed for a variety of surface processing applications. Deng et al. [15] proposed the optimization of plasma oxidation and abrasive polishing to increase the material removal rate (MRR) applied on substrate 4H-SiC using the plasma-assisted polishing (PAP) process. It showed that the plasma oxidation rate was significantly enhanced with a low water vapor content in He gas. Plasma process removes the material from the workpiece by creating low-temperature plasma inside the plasma chamber. Yadav et al. [3] investigated the atomistic material removal rate using a medium-pressure plasma process (MPPP) on a fused silica surface, and further material removal mechanism is discussed. Plasma polishing parameters, i.e., gas composition, pressure ratio, total pressure, and RF power, affecting the material removal rate of fused silica, have been presented. Field-emission scanning electron microscopy (FESEM) is an advanced imaging technique used to visualize the surface structure and topography of a sample at high resolution [16, 17]. Yadav et al. [18] investigated plasma-processed fused silica substrate, and it is characterized using FESEM and EDX, which depict the presence of silicon, oxygen, and fluorine on the processed substrate. This technique overcame all the drawbacks using traditional methods and gave a high-quality surface finish on optical parts [19]. The finishing of optical components using the traditional polishing method, more surface damage occurs, and it also produces some micro-cracks in the surface [20]. All those factors affect the performance of the optical components. These drawbacks happen while using traditional polishing methods due to physical contact between the workpiece and the tool [21]. To satisfy all those requirements, research in the field of optical polishing with soft material surfaces, a novel non-contact MPPP, has been developed for the finishing of optical components. It removes the sub-surface damage from the surface of optical component.

The present study conducts experiments to find the surface roughness of optical components. The novelty of the process arises from the fact that the current medium-pressure plasma polishing process is capable of polishing complex 3D profiles with cavities where neither a tool nor a beam can reach and can remove the material from the surface without any contamination deposition during processing. Ionized atoms in the plasma chamber help remove material from the fused silica surface. This paper presents the schematic diagram, an actual photograph of the developed experimental setup, and the principle of the plasma polishing process. Moreover, the surface roughness, morphology, and chemical composition of the processed surface before and after the plasma process using a 3D optical profiler, FESEM, and EDX, respectively. The research challenges and future scope have also been discussed.

2 Materials and Methods

The following section presents the plasma process principle, the plasma processing flow chart, and the schematic and actual photograph of the developed experimental setup.

2.1 Principle of Plasma Process

In the polishing process, the process and reactive gas with the optimal ratio are properly mixed and then admitted into the process chamber. The reaction gas is then ionized by radio-frequency (RF) power in the plasma chamber to generate high-density energy reactive radicals [22]. The plasma generates the radicals and ion

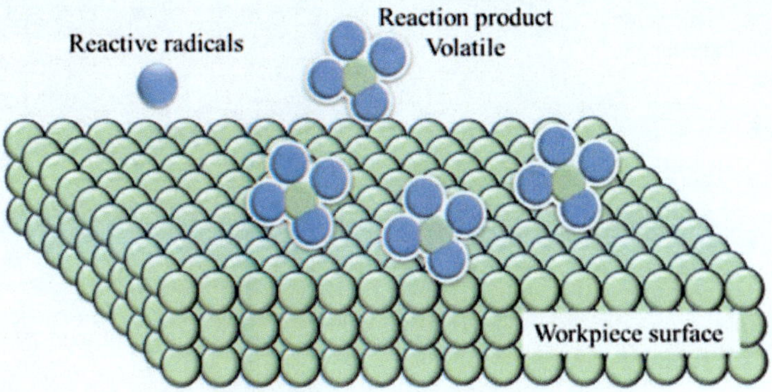

Fig. 1 Principle of plasma process

species which react with the surface of the workpiece as energetic radicals, ions, and electrons. The reactive radicals then interact chemically with the workpiece's surface atoms. These radicals make surface reactions with the fused silica, which makes effective atomic-scale material removal. Further, the volatile product, i.e., SiF_4, generated and removed the material from the surface, as shown in Fig. 1. This process eliminates the material from the surface of the workpiece by the radical and ion species. The probable reaction is illustrated in Eqs. (1) and (2) [3].

$$SiO_2 + 4F^* \rightarrow SiF_4 \uparrow + O_2 \uparrow \tag{1}$$

$$Si + 4F^* \rightarrow SiF_4 \uparrow \tag{2}$$

2.2 Experimental Setup

The medium-pressure plasma polishing setup consists of gas cylinders (He, O_2, and sulfur hexafluoride (SF_6)), mass flow meters (MFM, sccm: standard cubic centimeter per minute), a plasma chamber, a pump, and RF power. The gasses flow into the mass flow meter from the gas cylinder, and a separate controller is used to control the flow rate of the gasses and admitted into the plasma chamber. The gasses are allowed to enter into the chamber, and RF power is used to ionize the gases inside the chamber, leading to the production of plasma inside the chamber. The generation of plasma inside the process chamber during polishing is shown in Fig. 2.

The schematic diagram and actual photograph of the developed experimental setup of the medium-pressure plasma process are illustrated in Fig. 3a and b, respectively. The method achieved an ultra-smooth finish with a high material removal rate and good surface finish [18]. The gases are coming out from gas cylinders, and a mass flow meter is used to control the flow of gases. Further, the gases are filled in the

Fig. 2 Generation of plasma inside the chamber

plasma chamber with optimum ratio and ionized using RF power. The pump is used to create the vacuum inside the chamber just before the filling of gases, and also, it helps to exhaust generated gases during processing.

2.3 Flow Chart of Medium-Pressure Plasma Polishing Process

The machining stages in the medium-pressure plasma polishing process are organized in circular diagrams in a flow chart, as shown in Fig. 4. Initially, the process gases come from different gas cylinders and pass through mass flow meters. It controls the flow rate of the gases. The gases are then admitted into the plasma chamber at a particular partial pressure controlled by the flow rates of mass flow meters. RF power ionizes the gases into radicals, ions, atoms, and molecules. The radicals react with fused silica substrate and generate volatile silicon tetrafluoride (SiF_4) compounds.

3 Results and Discussion

The following section provides fused silica's surface roughness, morphology, and elemental composition before and after processing using a 3D optical profiler, FESEM, and EDX.

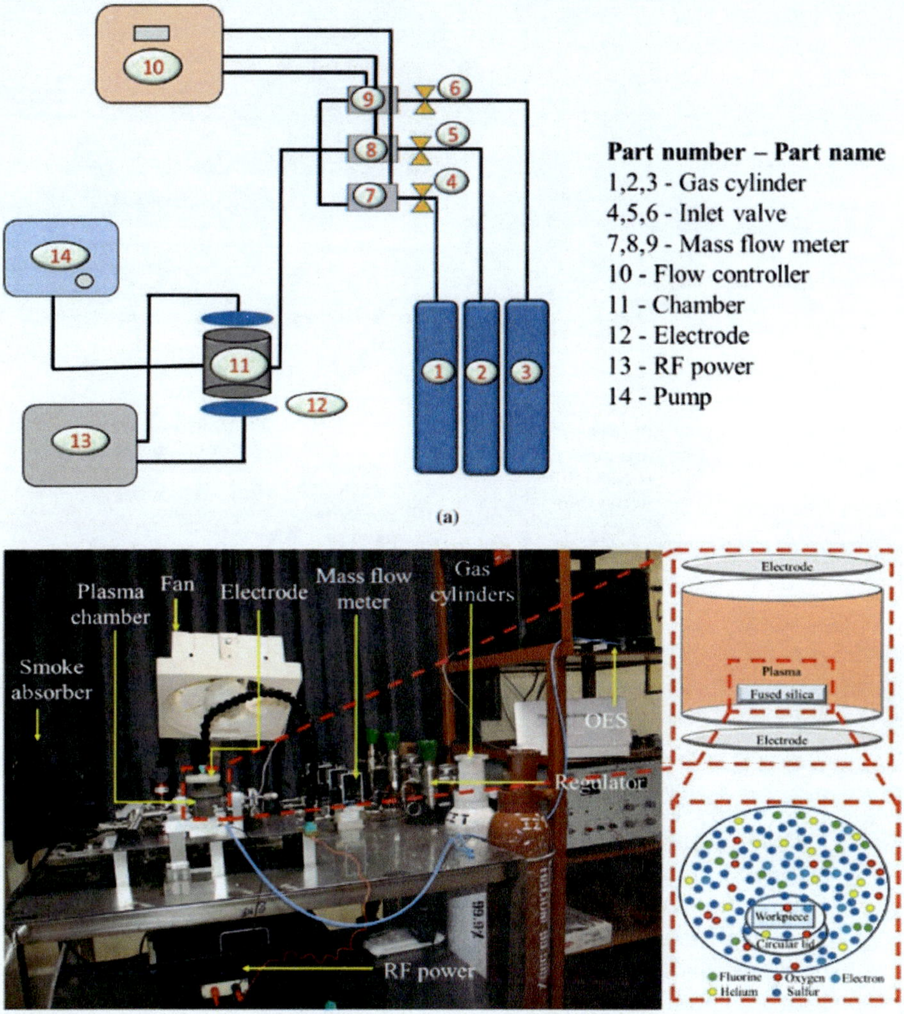

Fig. 3 Experimental setup of medium-pressure plasma polishing **a** schematic diagram and **b** actual photograph

3.1 Surface Finishing of Fused Silica Substrate

The experiments are carried out using He, O_2, and SF_6 plasma to investigate the surface roughness of fused silica. The process parameters, i.e., gas composition, pressure ratio, total pressure of plasma chamber, and RF power, have been chosen based on the previous literature survey and preliminary experiments. Higher discharge directly affects the surface profile in the plasma polishing process. Complete plasma discharge

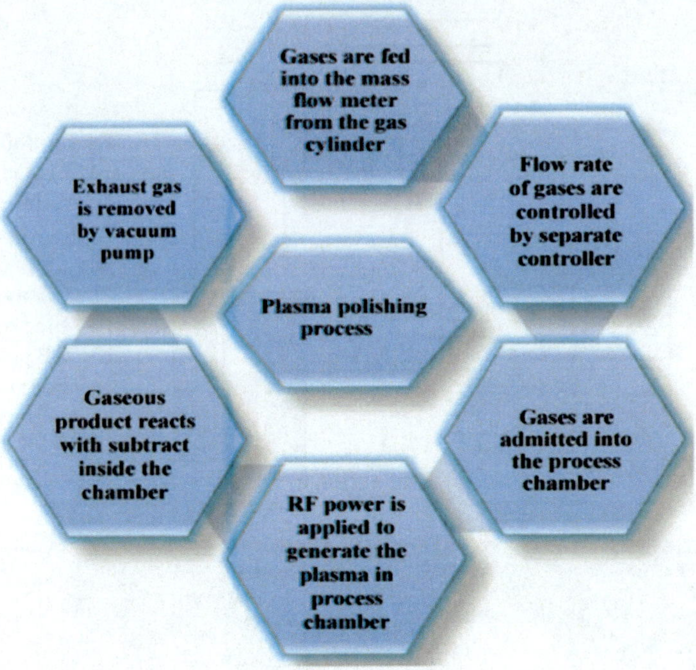

Fig. 4 Flow chart of the plasma polishing process

leads to uniform material removal from the substrate. The processing parameters of the plasma process are illustrated in Table 1.

Figure 5a and b shows 2D roughness profiles before and after plasma processing, respectively. The experiments have been performed on the process parameters, i.e., gas composition (He:($SF_6 + O_2$)) of 92:8, pressure ratio (SF_6/O_2) of 1:1, RF power of 80 W, and machining time of 55 min. The surface roughness is marginally increased from 0.35 μm to 0.37 μm after plasma processing. Before and after plasma processing, the surface finish of fused silica is measured by a 3D profiler. It is evident

Table 1 Processing parameters of plasma experiment

Parameters	Values
Gas composition (He: ($SF_6 + O_2$))	92:8
RF power	80 W
Pressure ratio (SF_6/O_2)	1:1
Flow rate of He	58 sccm
Flow rate of O_2	0.90 sccm
Flow rate of SF_6	1.40 sccm
Total pressure of plasma chamber	20 mbar
Machining time	55 min

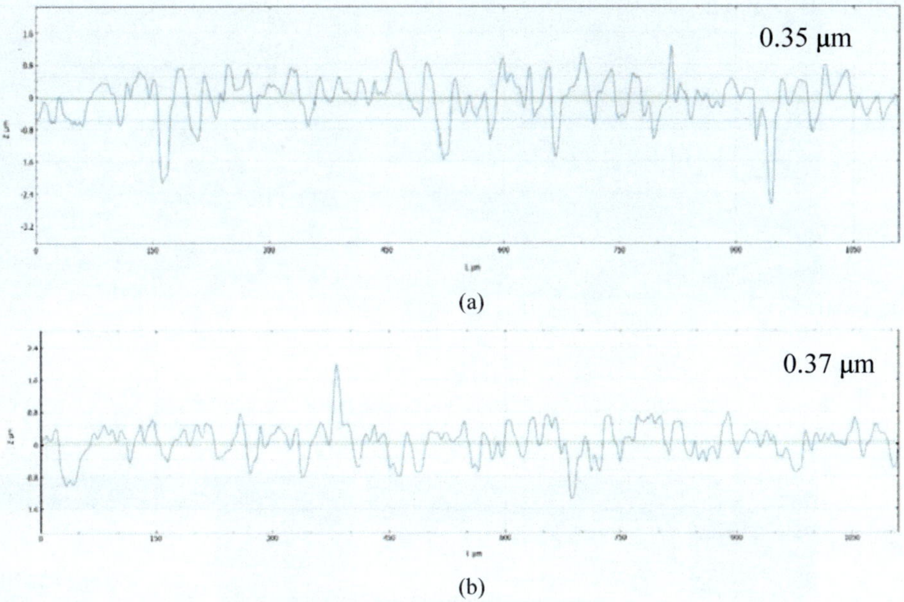

Fig. 5 Surface roughness of fused silica **a** before and **b** after medium-pressure plasma process

by the workpiece contrast that the substrate surface, after plasma processing, achieved a good surface finish without any surface defects and surface contamination. Three surface profiles are used to measure each sample's initial and final surface roughness values. After that, an average value of surface roughness is calculated.

3.2 *Microstructure and Chemical Composition Analysis*

FESEM provides the morphology of the substrate's surface. The surface morphology of fused silica before and after processing is illustrated in Fig. 6a and b, respectively. The micro-cracks and adhered materials have been detected on the original surface, as illustrated in Fig. 6a. The results showed that the cracks, etched pit, and deposition had been removed after the plasma process, as shown in Fig. 6b. EDX is used to analyze the elemental composition on the substrate surface. The EDX findings of fused silica before and after the plasma process are presented in Fig. 7a and b, respectively. It is obvious that before processing Si, O, and after the process, it consists of four elements such as Si, F, O, and C, detected on the surface of fused silica, where the majority of the oxygen and carbon elements come from atmospheric air. The presence of fluorine elements shows that the reaction occurred during processing.

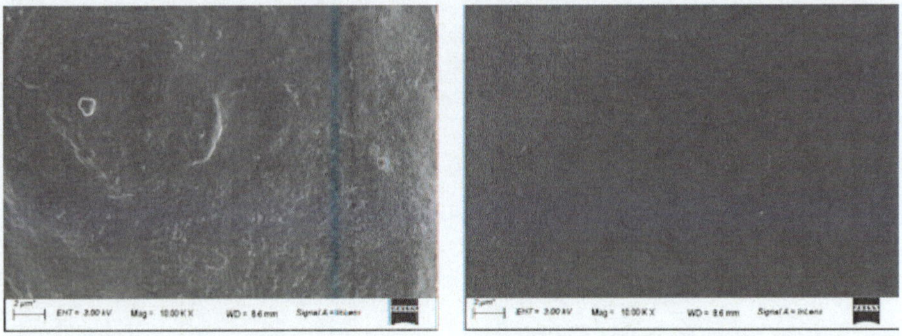

(a) (b)

Fig. 6 FESEM image of fused silica **a** before and **b** after medium-pressure plasma process

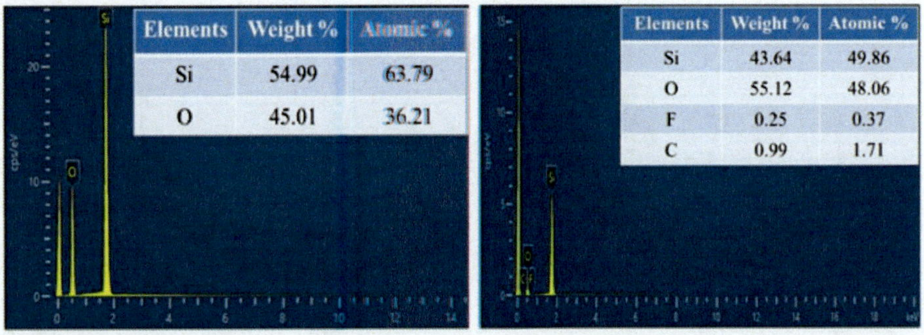

(a) (b)

Fig. 7 EDX image of fused silica **a** before and **b** after medium-pressure plasma process

4 Research Challenges

The current development of optics-related industries requires the extremely severe quality of the surface with less damage. This type of requirement is the present challenge of optic polishing technology. Using traditional processes always creates damage, such as micro-cracks and lattice disturbances, on the surface of the workpiece. The conventional polishing method is not suitable for functional brittle, hard materials like glass, ceramics, and crystal due to the individual characteristics of those materials. For the finishing of optical materials, most of the researchers used non-conventional plasma processes and utilized some common gases like sulfur hexafluoride (SF_6), helium (He), argon (Ar), and carbon tetrafluoride (CF_4) to polish optical materials such as fused silica, silicon, ceramics, and zerodur materials. Very few researchers have discussed polishing the critical complex and freeform optical surfaces. Several researchers have reported that the material removal rate of optical material is 5–32 mm^3/min using a plasma process. However, the efficiency of this process is low, and improvements in the surface roughness are not so much good.

During processing, it requires periodically cleaning the process chamber to make the finishing process more efficient. The plasma process is mainly utilized to improve the sub-surface defect of optical material.

5 Future Scope

The plasma process has a promising future and is expected to have several significant applications across various fields. The process parameters can explore in some higher range to get the optimum MRR and better surface finish. There are some potential areas where the plasma process can make a significant impact:

- Surface modification and coating: Medium-pressure plasma processes can be used for surface modification and deposition of functional coatings. With medium-pressure plasmas, treating a wide range of materials, including polymers, ceramics, metals, and composites, is possible. These processes can improve surface adhesion, enhance wear resistance, provide corrosion protection, and introduce specific functionalities like hydrophobicity or antibacterial properties.
- Electronics manufacturing: Medium-pressure plasma processes find extensive applications in the electronics manufacturing industry. They can be used for cleaning and activating surfaces prior to bonding or soldering processes, improving the adhesion and reliability of electronic components. Medium-pressure plasma can also be employed for the surface treatment of printed circuit boards (PCBs), enhancing their performance and ensuring better solderability.
- Semiconductor industry: The medium-pressure plasma process is well-suited for various semiconductor manufacturing steps. It can be utilized for precise etching of semiconductor materials, such as silicon, gallium arsenide, and other compound semiconductors. Medium-pressure plasma processes can also facilitate the deposition of thin films and patterned coatings used in integrated circuits and other electronic devices.

6 Conclusion

This article briefly discusses the optical substrates, application of optical materials, schematic and actual photograph of the experimental setup, and principle of the plasma polishing process. 3D optical profilometer, FESEM, and EDX are used to find out the surface roughness, morphology, and chemical composition of the processed surface are included. The following conclusion can be drawn:

- The surface roughness of fused silica has slightly increased from 0.35 μm to 0.37 μm after the treatment of the plasma process.
- FESEM results show that fused silica morphology has improved, such as cracks, etched pits, and holes after the plasma process.

- EDX results show that the elements Si and O are presented on the initial surface, and F, C, Si, and O are observed on the processed surface after the plasma process. The presence of fluorine elements shows the reaction on the fused silica's surface.

Acknowledgements We acknowledge the Science and Engineering Research Board, New Delhi, India, for their financial support for project No. ECR/2018/002801 entitled "Design and development of a novel plasma processing set up for uniform nanopolishing of prism and any freeform surfaces of fused silica."

References

1. Yadav HNS, Kumar M, Kumar A, Das M (2021) Plasma polishing processes applied on optical materials: a review. J. Micromanuf. 6(1):251659842110388
2. Enni K, Dev DSD, Das M (2020) Induction of conditioning gas and its optimization in nonconventional plasma machining process of fused silica. In: Lecture notes in mechanical engineering, Singapore, pp 549–559
3. Narayan H et al (2023) Investigation of MRR and surface characterization using plasma process. Mater Manuf Process 1–13
4. Yadav HNS, Kumar M, Das M (2022) Fundamentals of plasma polishing. In: Jain VK (ed) Advanced machining science. CRC Press, Boca Raton, pp 203–228
5. Yadav HNS, Kumar M, Kumar A, Das M (2021) COMSOL simulation of microwave plasma polishing on different surfaces. Mater Today Proc 45:4803–4809
6. Dev DSD, Krishna E, Das M (2016) A novel plasma-assisted atomistic surface finishing on freeform surfaces of fused silica. Int J Precis Technol 6(3):262
7. Li R, Li Y, Deng H (2022) Plasma-induced atom migration manufacturing of fused silica. Precis Eng 76:305–313
8. Dev DSD, Enni K, Das M (2018) Novel finishing process development for precision complex-shaped hemispherical shell by bulk plasma processing. In: Precision product-process design and optimization, pp 313–335
9. Arnold T, Böhm G (2012) Application of atmospheric plasma jet machining (PJM) for effective surface figuring of SiC. Precis Eng 36(4):546–553
10. Paetzelt H, Böhm G, Arnold T (2013) Plasma jet polishing of rough fused silica surfaces. In: Proceedings of the 13th international conference of the european society for precision engineering and nanotechnology, vol 2, pp 19–22
11. Jin H, Xin Q, Li N, Jin J, Wang B, Yao Y (2013) The morphology and chemistry evolution of fused silica surface after Ar/CF 4 atmospheric pressure plasma processing. Appl Surf Sci 286:405–411
12. Gerhard C, Weihs T, Tasche D, Brückner S, Wieneke S, Viöl W (2013) Atmospheric pressure plasma treatment of fused silica, related surface and near-surface effects and applications. Plasma Chem Plasma Process 33(5):895–905
13. Cui Z et al. (2023) SF6 degradation in a γ-Al2O3 packed DBD system: effects of hydration, reactive gases and plasma-induced surface charges. Plasma Chem Plasma Process 1–22
14. Graves DB (1994) Plasma processing. IEEE Trans Plasma Sci 22(1):31–42
15. Deng H, Monna K, Tabata T, Endo K, Yamamura K (2014) Optimization of the plasma oxidation and abrasive polishing processes in plasma-assisted polishing for highly effective planarization of 4H-SiC. CIRP Ann 63(1):529–532
16. Singh Yadav HN, Das AK, Das M (2022) Synthesis of Tungsten carbide nanoparticles in different dielectric through μ-EDM. Adv Mater Process Technol 9(1):1–9

17. Yadav HNS, Bishwakarma H, Kumar N, Kumar S, Singh PK, Mohanty S, Das (2019) Production of tungsten carbide nanoparticles through Micro-EDM and its characterization. Mater Today Proc 18:1192–1197
18. Narayan H, Yadav S, Das M (2023) Parametric optimisation of plasma polishing process using response surface methodology. Surf Eng 39(2):1–14
19. Tendero C, Tixier C, Tristant P, Desmaison J, Leprince P (2006) Atmospheric pressure plasmas: a review. Spectrochimica Acta Part B Atomic Spectroscopy 61(1):2–30 (Elsevier)
20. Zhang J, Wang B, Dong S (2008) Application of atmospheric pressure plasma polishing method in machining of silicon ultra-smooth surfaces. Front Electr Electron Eng China 3(4):480–487
21. Xie M, Pan Y, An Z, Huang S, Dong M (2022) Review on surface polishing methods of optical parts. Adv Mater Sci Eng 2022(1):1–30
22. Dev DSD, Krishna E, Das M (2019) Development of a non-contact plasma processing technique to mitigate chemical network defects of fused silica with life enhancement of He-Ne laser device. Opt Laser Technol 113:289–302

Additive Manufacturing with Interdisciplinary Applications

Mohammad Taufik and Prashant K. Jain

Abstract Additive manufacturing (AM) becomes the successful commercial technology because of the interdisciplinary applications of this technology in different sectors. This review paper categorises the interdisciplinary applications of AM into three different sectors as prototyping, rapid tooling and direct part production. Each application sectors provide reviews on how normally AM processes have been adopted, albeit how and why they are utilised has also been presented. In this paper, sufficiently novel and important observations have been presented which include AM fabricated metamaterial, microelectrode arrays, micro-channel, mesoscale burner, marine parts, monolith structures, biofuel microstructured and assembly with photovoltaic. In this way, the present approach provides significant insights into different additive manufacturing application sectors. In addition, to provide an overall view, interdisciplinary applications listed in the current study are sufficiently adequate for a review in terms of number of research works considered from the literature and various resources available in the public domain. The current research also shows that the AM can provide future scope in different fields such as cold spray, direct fabrication of electromechanical devices on the macroscale, radio-frequency structures and similar applications.

Keywords Additive manufacturing · 3D printing · Application sectors

M. Taufik (✉)
Maulana Azad National Institute of Technology Bhopal, Bhopal, Madhya Pradesh 462003, India
e-mail: mohammad.taufik@manit.ac.in

P. K. Jain
PDPM Indian Institute of Information Technology, Design and Manufacturing Jabalpur, Jabalpur, Madhya Pradesh 482005, India

N. Kumar et al. (eds.), *Advances in Materials and Agile Manufacturing*,
Lecture Notes in Mechanical Engineering,
https://doi.org/10.1007/978-981-99-6601-1_17

1 Introduction

Unlike to subtractive process, such as drilling, milling and similar machining processes. Additive manufacturing (AM) process fabricates solid objects by joining material layer upon layer from 3D CAD model. ISO/ASTM 2016 provides seven categorisations for the AM processes, namely directed energy deposition (DED), powder bed fusion (PBF), vat photopolymerization, sheet lamination (SL), binder jetting (BJ), material extrusion and material jetting (MJ). These AM processes differ with reference to their mechanism of energy demands, layer thickness or beam width (e.g. resolution), volume built per time (e.g. build rate), material compatibility, fusion and other attributes. In spite of all these variations, these processes can be summarised in one place based on the resolution and productivity of each AM processes as shown in Fig. 1 [1].

AM processes are working with various materials such as metal, plastic and composites, without wasting material at large scale unlike the conventional manufacturing processes. It is frequently believed that AM offers "design freedom" implication that the adding of complex features does not effects AM part fabrication capability. To assess this theory in the current AM marketplace, conventional assemblies can be manufactured in a single complex structure through AM processes. Moreover, AM built complex parts do not require significant amount of post-processing. Further, the benefit of the AM process is that unlike the conventional CNC machining process part complexity increases the cost of production. However, in case of increase of part complexity in case of AM process cost does not increase, besides it remains relatively invariant [2] or decreases as shown in Fig. 2 [1]. Additional key application aspect of the AM process is that it is naturally very auspicious. As a result of all above-mentioned benefits of AM processes, its application as well as market share is increasing rapidly, spreading into various sectors, such as prototyping, rapid tooling

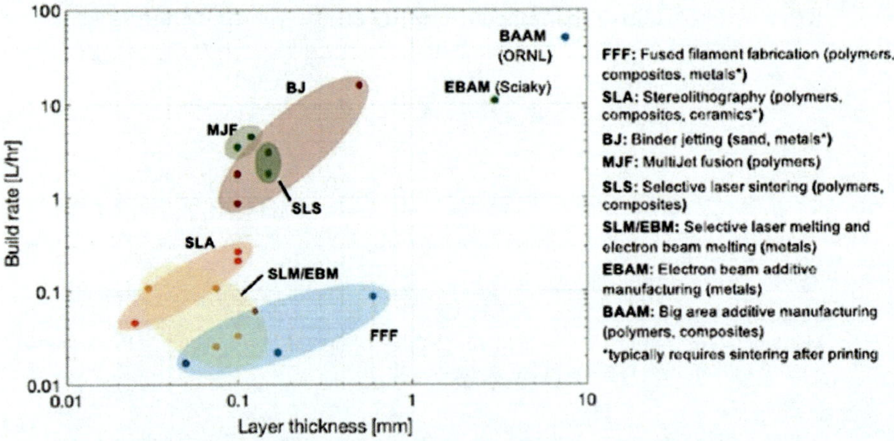

Fig. 1 Representation of different AM processes based on their productivity and resolution

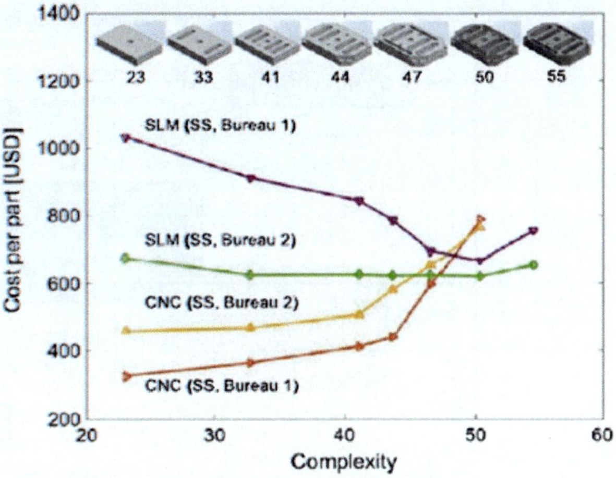

Fig. 2 Cost versus complexity-based comparison between AM and CNC process

and direct part production, and is anticipated that this substantial progression will remain for the next some years.

2 Application Sectors of AM

This section summarises applications of AM process in the current scenario. The current applications of AM processes are used extensively for prototyping, rapid tooling and direct part production. Thus, three different types of the applications of AM process have been presented in this section. These studies will also refer to show the effect of part surface quality on AM applications.

2.1 *Application in Prototype Development*

There are many types and classes of physical prototypes, but their main purpose is to minimise risk throughout the product development phase [3]. The objective of AM prototyping is to fabricate quickly any complex-shaped three-dimensional part from CAD data [3]. Some of the specific applications of prototyping follow engineering analysis and planning, design, concept models, art models, medical models, architectural models, form, fit and function testing and validation, assembly models and styling, ergonomic studies. Further, to explore the AM applications in computational research and its short- as well as long-term future goals which were suggested and summarised by Smith et al. [4]. Smith et al. [4] has suggested a linkage relationship between process–structure and structure–property along with the structure and direction, which is presented in Fig. 3. However, for substantial

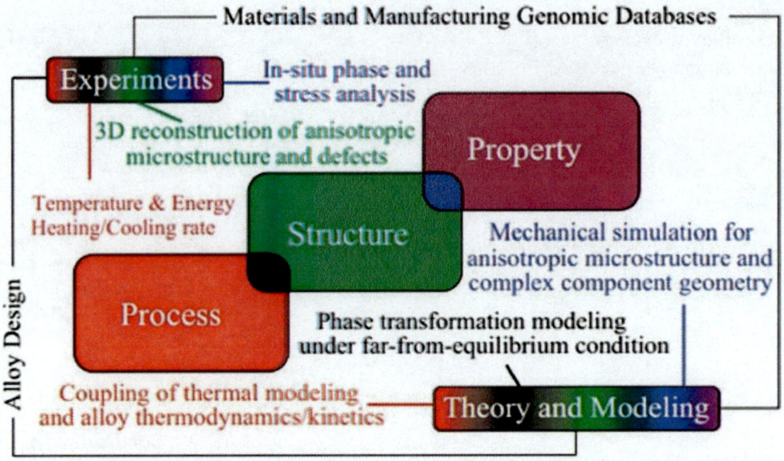

Fig. 3 Future goals of computational modelling for AM applications

enhancement of AM application corresponding experiments must be calibrate and/or validate theoretical modelling.

Recently, unlike the conventional AM functional prototyping applications were outer shape of objects engineered, AM prototyping also engineers internal microstructures such as materials create along with hard as well as soft regions and it is known as metamaterial [5, 6].

Ion et al. [5, 6] have demonstrated mechanical function of AM fabricated metamaterial objects. These AM printed metamaterial objects consists of well-defined single block of material cells mechanism which helps them to complete macroscopic movement in different case study such as door latch, digital door latch, pliers, pantograph, Jansen walker and switch as shown in Fig. 4.

Figure 5a, b demonstrates the application of AM in art. Complex and intricate artworks can be easily fabricated using AM processes as there is no limit on printing complex parts. Figure 5b shows a display case along with a splash spanning object as shown in Fig. 5a which is fabricated by AM. In these cases, surface quality plays an important role which will improve aesthetical characteristics.

Nowadays, architectural models and engineering analysis models are frequently produced by AM. Applications of prototyping in architecture started in the year 2000. Gibson et al. [8] discussed the application of AM for architectural modelling and its limit. Figure 5c shows the example of models made using AM. They discussed two different critical requirements of architects. According to their discussion, the first requirement is related to the separation of the various sectors and second is related to various materials in models. Gibson et al. [8] suggested that the RP technology can be used more creatively than usual by combining models made with different materials and processes.

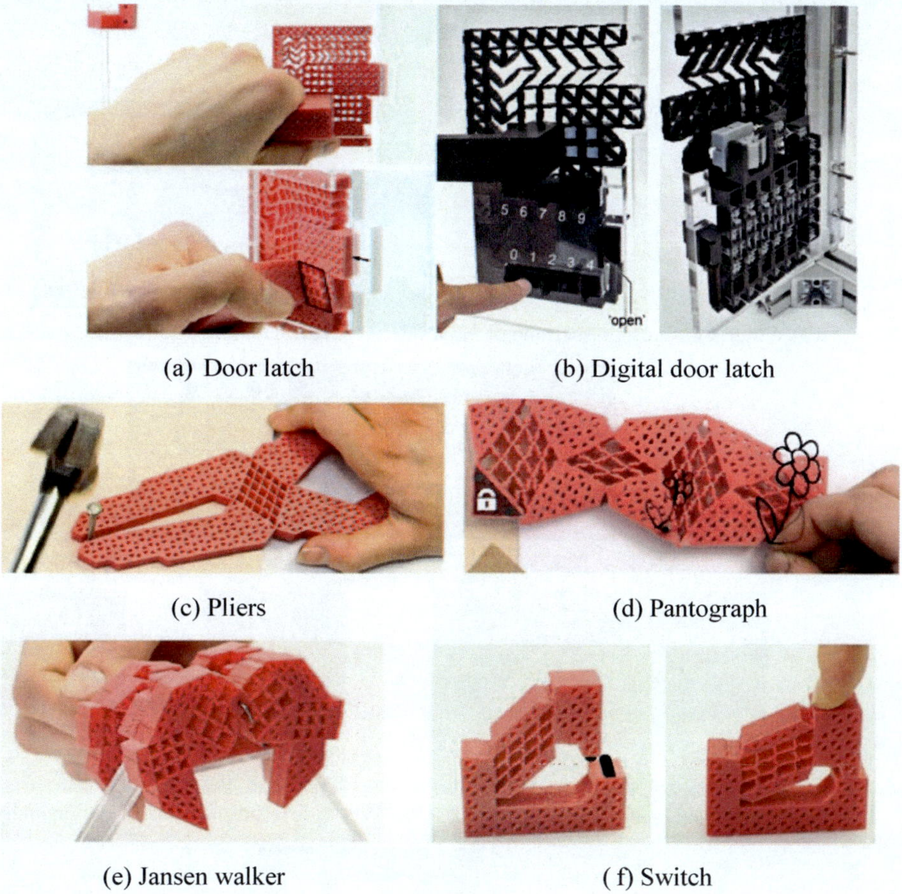

(a) Door latch (b) Digital door latch

(c) Pliers (d) Pantograph

(e) Jansen walker (f) Switch

Fig. 4 AM fabricated metamaterial mechanisms

Some literature has been observed to support RP application for engineering analysis models. As engineering analysis is a vast domain, the four typical applications of engineering analysis, as shown in Fig. 5d–f, are provided as follows: flow pattern visualisation, thermoelastic tension as well as photoelastic stress analysis and fabrication of wind-tunnel test models.

Nowadays, AM prototype is being used for assembly process planning. It offers numerous advantages in different application sectors and in the customised uses [9]. These advantages are usually missing in other methods like the computer model. Ahmad et al. [9] categorised the core issues such as shrinkage and scaling for assembly process planning using RP technology. They presented detailed guidelines for assembly and training processes using RP. In their work, assembly case study of a reciprocating gizmo mechanism has been taken. Figure 6 shows a typical case study which is used for the RP for assembly training and validation purposes. In this

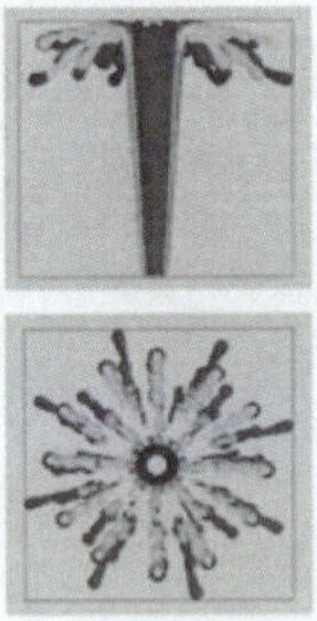

(a) Water splash cross-sections of 3D model

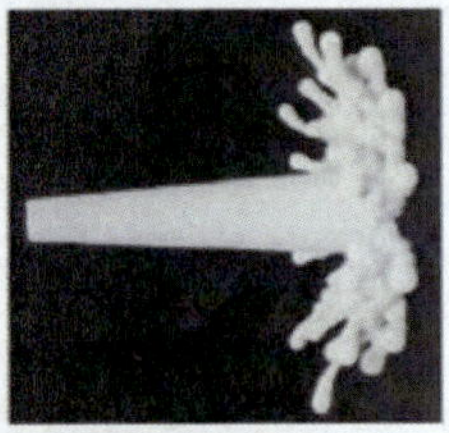

(b) SLS fabricated water splash model

(c) Space frame construction

(d) Visualisation of flow patterns

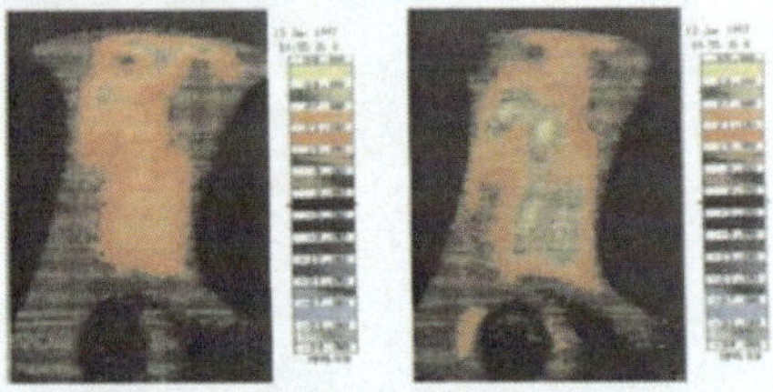

(e) Thermo elastic tension analysis

(f)Photoelastic stress analysis

Fig. 5 Applications of RP in **a**, **b** Art, **c** Architecture and **d**–**f** Engineering analysis [7]

case, any surface irregularities or poor surface quality issue significantly influences parts functionality or assembly.

Low-cost AM/3D printers along with friendly CAD applications facilities have made inroads into the classroom. Students around the globe in different schools and colleges are able to design and produce prototypes which they can observe and study for educational purposes. The use of AM process allows students to enhance their

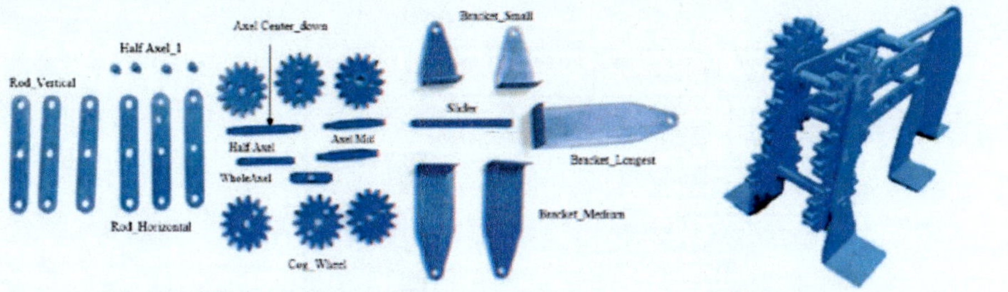

(a) AM fabricated assembly components (b) Reciprocating gizmo mechanism

Fig. 6 AM prototypes for assembly training and validation [9]

learning abilities. Kroll and Artzi [10], for example, have used 3D-printed wind-tunnel models for improving learning skills of aerospace engineering students. They also found that the application of AM in prototype development is very economical for educational purposes.

Different studies on the impact of AM on engineering education have been performed by Minetola et al. [11]. Different study groups of students worked at Politecnico di Torino University, Italy. The projects on the AM of a downhill bike were also demonstrated. Figure 7 shows the FDM fabricated braking systems. The AM costs for the downhill bike are also estimated for the educational purposes which include C_{MAT} (material cost), C_{PRE} (pre-processing cost), C_{BUILD} (building cost) and C_{POST} (post-processing cost) and presented as follows:

$$C_{\mathrm{TOT}} = C_{\mathrm{MAT}} + C_{\mathrm{PRE}} + C_{\mathrm{BUILD}} + C_{\mathrm{POST}}. \tag{1}$$

They found highly positive feedbacks from students and suggested that the education of AM will empower people to extend products lifecycle and durability. In this

(a) Rear wheel structure (b) Detail of the transmission

Fig. 7 AM prototype of a downhill bike [11]

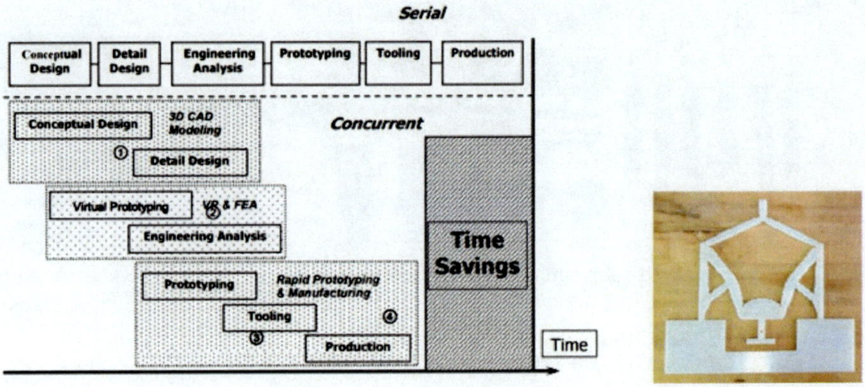

(a) Time compression engineering (b) Compliant mechanism

Fig. 8 AM/3D printing in rapid product development [7, 12]

work, a pre-processing cost is included which shows it possible to reduce total cost if surface quality of fabricated prototype is sufficient.

Moreover, need for fast-paced market and consumer goods demands rapid product development cycles. This led to AM being used in product design process for quick conceptualisation of idea into a physical prototype and design iteration. Proof-of-concept models can be easily made using AM processes. Pham and Dimov [7] graphically presented the benefits of product development (PD) process. The application of AM in prototype development has been used for time compression engineering to decrease the time to market. It can be observed from Fig. 8a that the prototyping provides significant time compression in product development.

Meisel et al. [12] have presented a case study for the design of optimised, multi-material compliant mechanism. With the added advantages of multi-material polyjet printing, such as geometric design freedoms and multi-material capability, compliant mechanisms were fabricated easily with optimised design as shown in Fig. 8b.

The above-presented different applications of AM prototypes provide motivation to study surface quality. Providing sufficient information about surface quality in prototypes will benefit prototype applications with a greater impact.

2.2 Application in Rapid Tooling

Rapid tooling is a unique application of AM for pre-production models. The term rapid tooling is normally utilised to define a process that either uses an AM built pat as a pattern to create moulds or uses the AM technique to fabricate tools directly for a limited volume of prototypes [3]. RT also refers to mould cavities that are either indirectly or directly fabricated using the AM process.

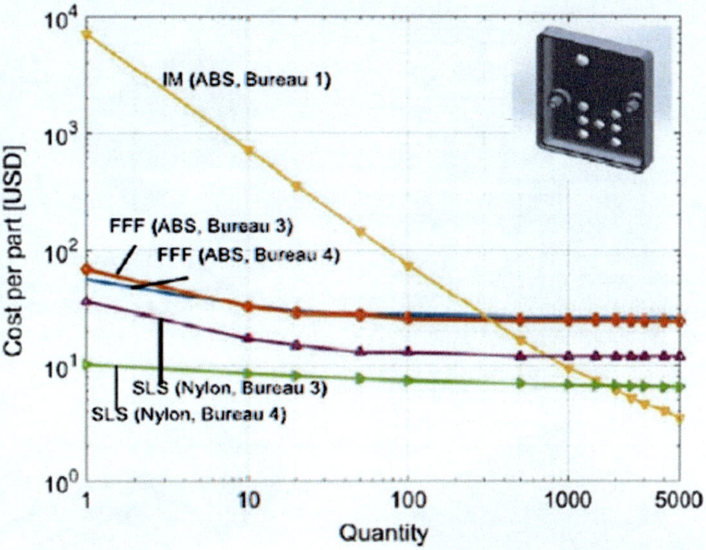

Fig. 9 Comparison among polymer components for cost versus quantity

RT is very important application of AM as it helps in enhancing market time rapidly, dropping investment and producing greater quality moulds and patterns. As a result of this nowadays there are many examples associated with RT as shown in Fig. 10. For tooling requirements, polymer-based AM processes overcoming the fixed cost obstacle as it delivers a nearly uniform cost-quantity connection. However, based on utilisation of AM machines its process parameters as well as corresponding post-treatment techniques and its factors may influence precise cost-volume trend for AM [1] as shown in Fig. 9. Rapid tooling process such as laser engineered net shaping (LENS) (please refer Fig. 11a) can be utilised for short-run manufacturing, repairing, rebuilding and functional prototyping. The complex features of a piece of jewellery can be precisely manufactured with RP pattern (please refer Fig. 11b).

The application of RT usually classified into two major types, namely indirect RT and direct RT. Indirect RT shows the fabrication of pattern by RP technique, which is further used as a model for die and mould making. Direct RT shows the fabrication of moulding and casting inserts directly from 3D CAD models. Indirect method for RT can be classified as room temperature vulcanizing silicone rubber moulds, spray metal tooling, sprayed steel, cast aluminium and zinc kirksite tooling, three-dimensional keltool, vacuum casting, reaction injection moulding, spin casting, cast resin tooling, rapid solidification process, plaster moulds, electroforming and investment cast tooling.

Mun et al. [21] presented a typical example of indirect rapid tooling processes which is used to produce ceramic mould by using sacrificial pattern as shown in Fig. 12a, b. In this process, two master patterns were assembled with a gating system which was attached to a system of gates, runners, sprue and pouring cup as shown in Fig. 12c. Ceramic slurry containing colloidal silica and zircon was used to dip

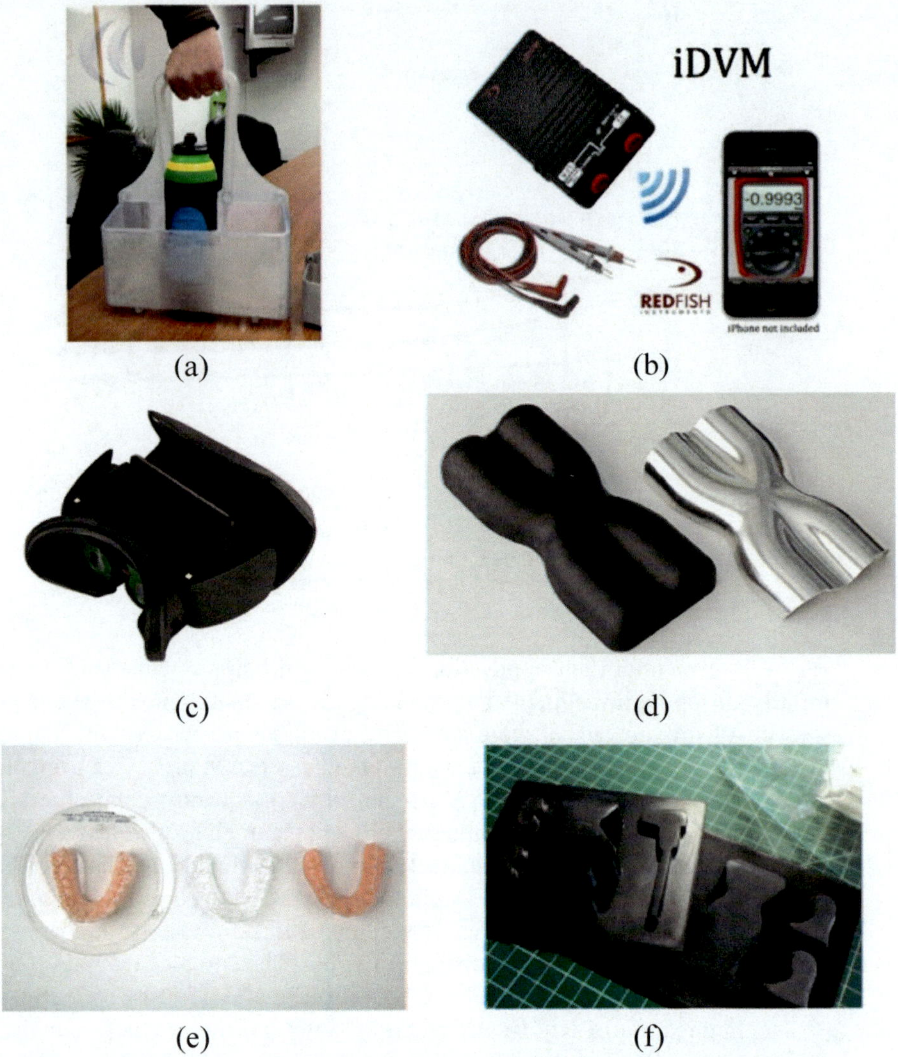

Fig. 10 AM application in rapid tooling for **a** water carrier: FIFA 2014 [13], **b** digital multimeter [14], **c** housings and the exterior cover [15], **d** hydroformed Diamond [13], **e** thermoformed aligners [16], **f** vacuum mould [17], **g** injection moulds [18], **h** paper pulp mould [19], **i** jigs and fixtures [20] and **j** investment casting [13]

pattern assembly as shown in Fig. 12d. Fine-grained zircon was used further for coating purpose after slurry was completely drained. Then pattern assembly was dry at room temperature as shown in Fig. 12e, after finishing of dipping and drying cycles dewaxing was completed as shown in Fig. 12f. The master pattern assembly was evacuated from the ceramic mould for forming the tool cavity. The ceramic mould was then moved to a vacuum chamber where it was filled by molten aluminium alloy

(g) (h)

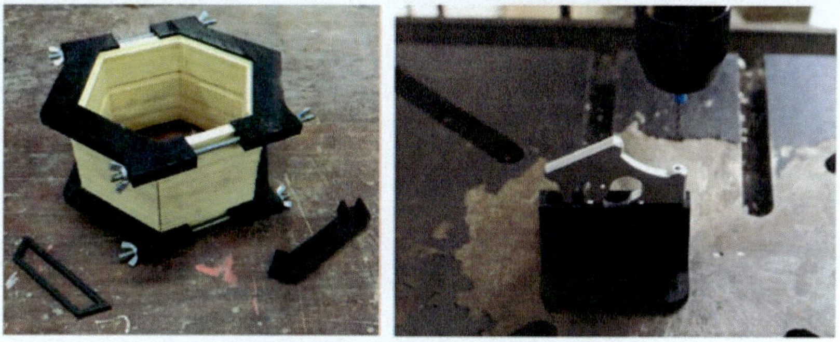

(i)

(j)

Fig. 10 (continued)

(AC4C). The material was poured under a vacuum condition to avoid any air-induced void asperities in final cast part as shown in Fig. 12g. After casting the material into the ceramic mould, it was solidified and cooled for the removal of ceramic mould which surrounded the casting as shown in Fig. 12h. Then casting part was used as a final rapid tool as shown in Fig. 12i which can be utilised further to produce multiple parts. But, due to the presence of one intermediate replication process in indirect RT process, tool has low accuracy, strength and requires higher build time. However, indirect RT process shows a better alternative to utilise AM process for the fabrication of tool to make few hundred parts. On the other hand, in direct RT process, when this one intermediate replication process is eliminated, the accuracy issues and build time decreases sharply. Direct method for RT can be classified as: direct aces injection moulding tooling, copper polyamide tooling, selective laser sintering rapid

(a) Parts produced by LENS and SL

(b) Ring produced by RP pattern

Fig. 11 AM application in rapid tooling [7]

steel, laminated tooling, laser engineered net shaping, topographic shape formation, direct metal tooling using 3DP™, controlled metal build up and prometal. Direct methods for RT cab also be classified into two types by the use of binder-coated powder approach and binderless material approach.

Petrovic et al. [22] presented the use of AM in mould making. In a simple mould fabrication process, a series of straight cooling channels are provided to provide adequate cooling. Moreover, in a mould-making process, warping, contraction and overall quality of part depends on the cooling. For example, in a simple mould fabrication process, if the direction of cooling channels is straight as shown in Fig. 13a, then it is not effective for homogeneous heat transfer. In other words, it does not provide the adaptive cooling to the mould's walls. On the other hand, in an AM fabricated cooling channels as shown in Fig. 13b, free-form features are provided so that it provides homogeneous heat transfer which helps to improve the quality of injected parts. Therefore, AM enables adaptive cooling and the injection cycle time reduces sharply. As compared to the conventional process, AM can also make mould inserts over manufactured mould blocks saving time and cost. Typical hybrid moulds are generally fabricated through electron beam or laser-based AM processes.

A further example of RT includes direct shell production casting (DSPC). In a DSPC, without any intermediate steps, it builds ceramic moulds from CAD file. This direct RT process has been promoted by Soligen Inc. and also known as 3DP™ process. To illustrate DSPC process, different steps are presented in Fig. 14. Further typical examples of 3DP™ process are shown in Fig. 15.

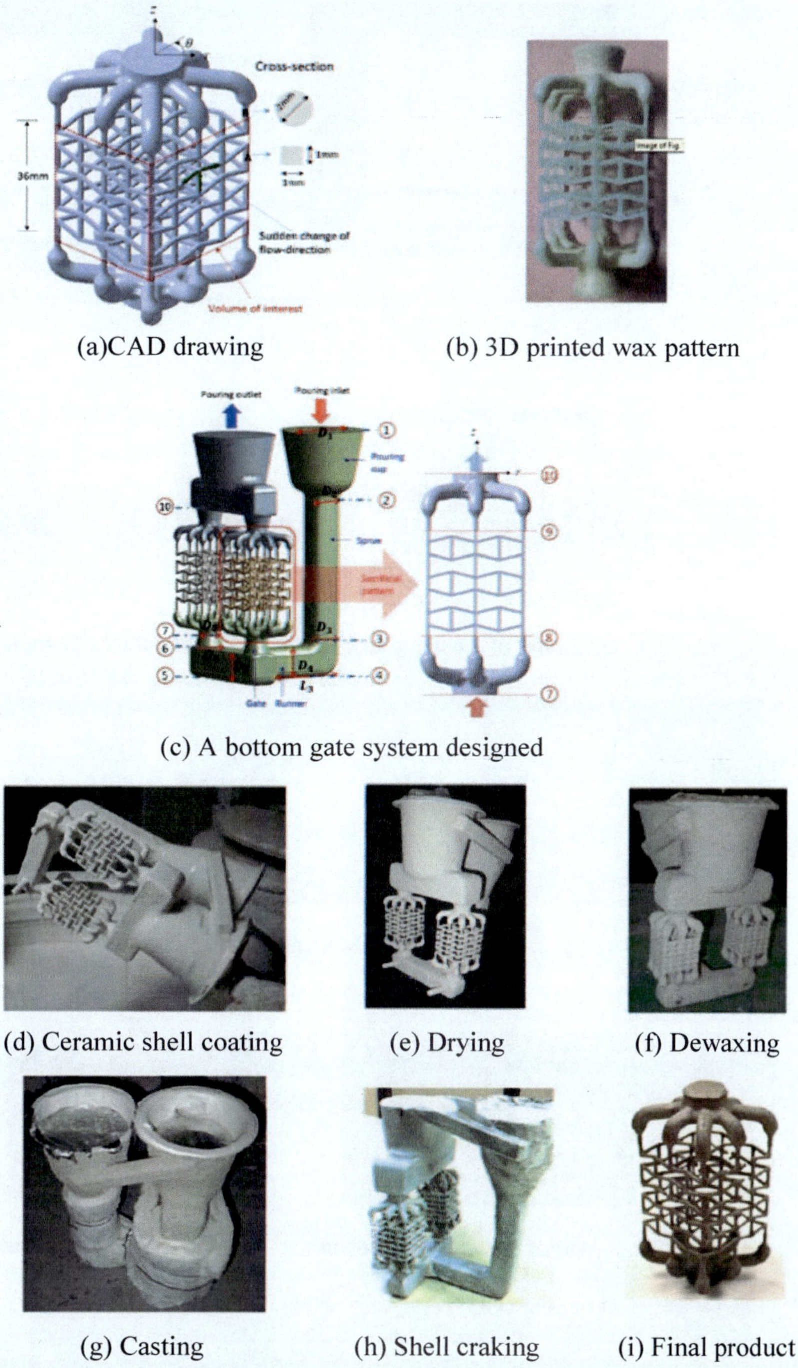

Fig. 12 Investment casting of 3D-printed cellular patterns [21]

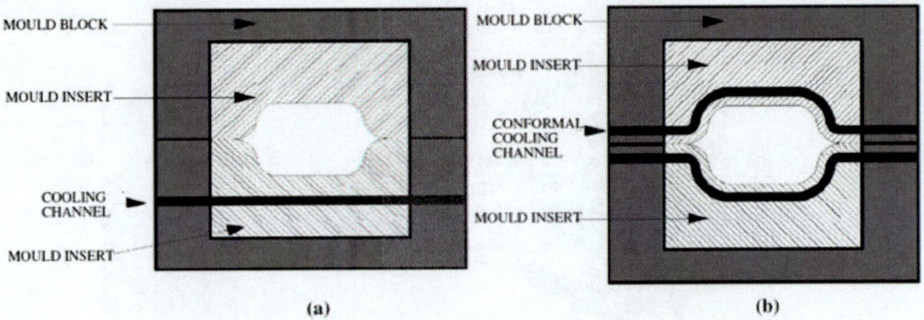

Fig. 13 **a** Conventional and **b** Conformal cooling channels [22]

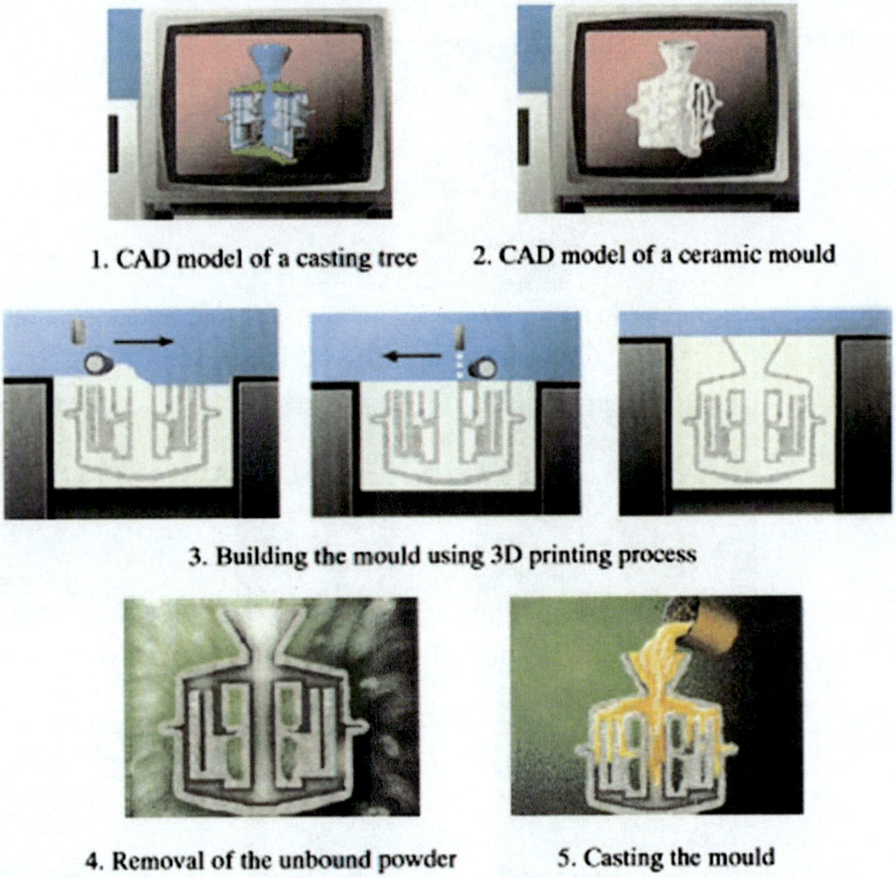

Fig. 14 Different steps of DSPC [7]

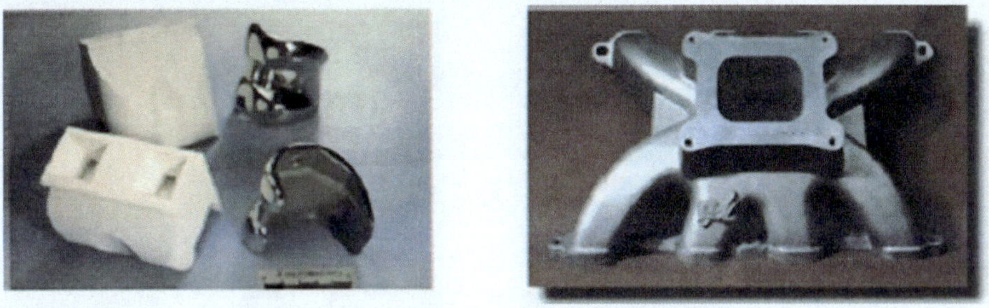

Fig. 15 Examples of 3DP™ process [7]

2.3 *Application in Direct Part Manufacturing*

In this section, some important applications of AM as a process to manufacture direct parts are presented. Several works have been recognised for the direct customised application of AM parts in medical. In general, the biggest area of AM processes for customised medical applications include orthoses and prostheses implant/replacement, medical devices, tissue engineering and scaffoldings as shown in Fig. 16.

In last three decades, AM has been used to print patient-specific orthoses and prostheses (O and P) for medical purposes. The most common types of O and P include prosthetic sockets, ankle–foot orthoses (AFOs) and foot orthoses (FOs). Figure 16a, b show the comparative examples of ankle–foot orthoses (AFO) and foot orthoses (FO) which were fabricated through traditional and AM process. AM provides more suitable conditions for direct part fabrication. For the restoration of tissues, scaffolds were usually printed through AM processes. This is because it is recognised that AM has the ability to fabricate the internal structure efficiently. A case study on the reconstruction of a mandibular implant for the fabrication of porous structures in a controlled manner for tissue engineering application is also presented in Fig. 16c, d. Moreover, AM is frequently used for the fabrication of medical devices, and hence, it increases the demand of surgical guidance aids.

Researchers are working in this direction, and they investigated the different uses of AM processes in surgical guides. Typical observations were documented through literature review. The major area of fabrication of medical devices includes hearing aid and dental industries implants. The significant ones are improvement in implant placement, surgical guides which are analogous to the optical tracking system and for keeping artificial replacement teeth (ART) in the oral cavity removable partial dental framework. Figure 16e–g shows the application of SL surgical guides to accurate dental implants placement. Moreover, part surface quality can play a major role in biomedical application area because the functionality of implant/replacement and scaffold surface topography are very critical in biomedical applications.

AM processes are also used to directly fabricate high-performance microelectrode arrays as shown in Fig. 17a. With the added advantages of AM process, it is easy to build microelectrode arrays in very quick operational time along with

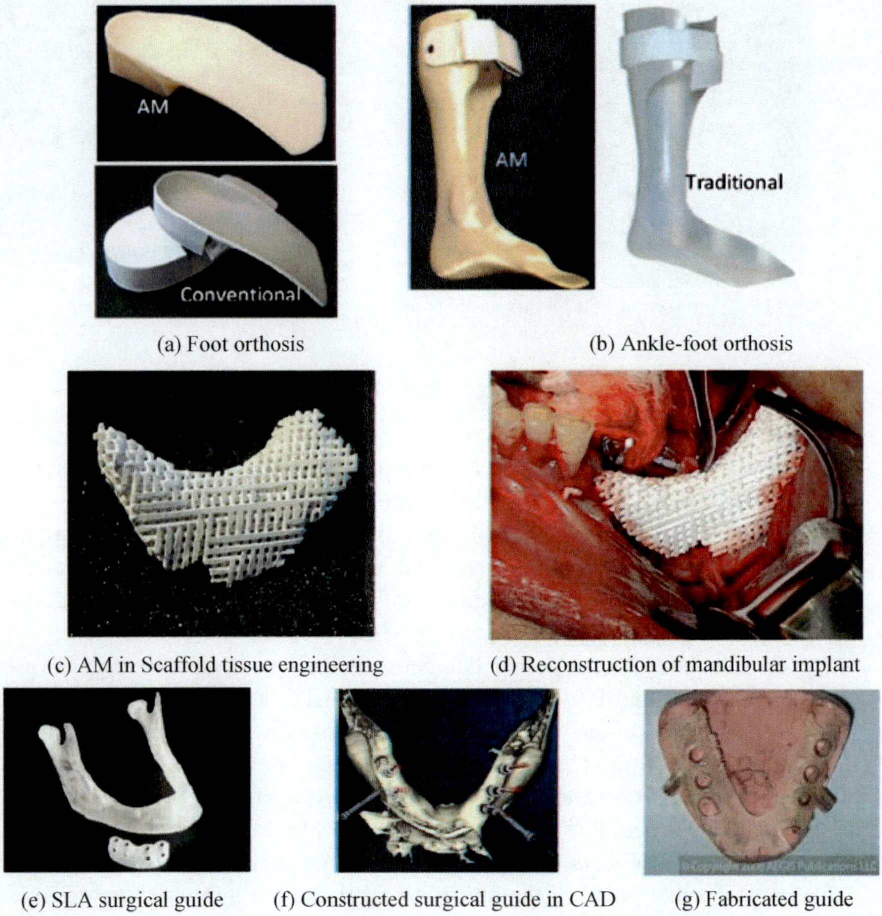

(a) Foot orthosis (b) Ankle-foot orthosis

(c) AM in Scaffold tissue engineering (d) Reconstruction of mandibular implant

(e) SLA surgical guide (f) Constructed surgical guide in CAD (g) Fabricated guide

Fig. 16 AM produced customised medical implants [23, 24]

low-cost and environmentally friendly conditions. However, AM surfaces have poor surface quality due to layer upon layer fabrication technique, which may contribute to variation in current per unit area in AM printed microelectrode arrays.

The AM printed microelectrode arrays (MA) can be used directly as an electrochemical platform. It can also be utilised for direct applications, namely biofuel cells, biosensors and other electrochemical devices. In this direction, Yang et al. [25] provide a novel work. In their work, 3D printing was performed with aerosol jet (AJ) technology for the fabrication of silver microelectrode arrays. Fabrications were performed with three different electrodes spacing. A schematic presentation for microelectrode arrays fabrication process is shown in Fig. 17. In their work, UV light source was used to instantaneously cure polymer ink during the fabrication process. AutoCAD modelling software was used to make polymer traces. To fulfil the need of sensing regions creation at different segments, polymer traces were fabricated with

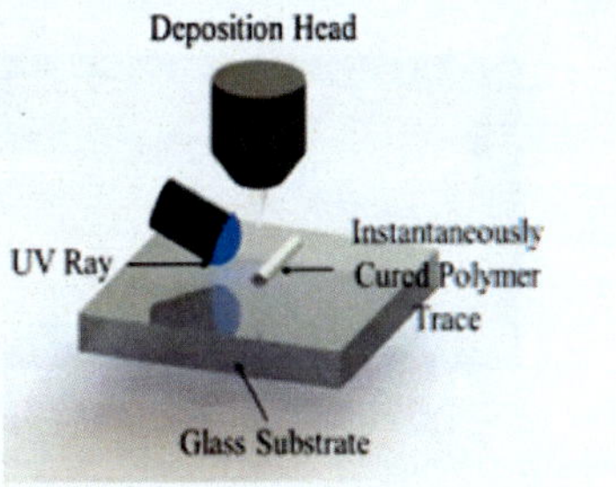

(a) Microelectrode arrays fabrication by AJ printer

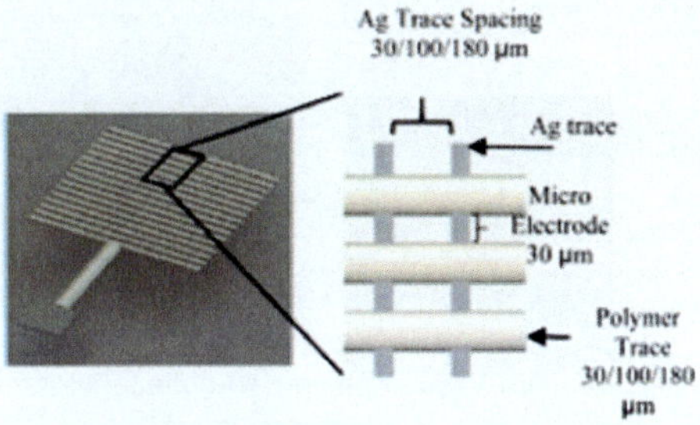

(b) Schematic of microelectrode array

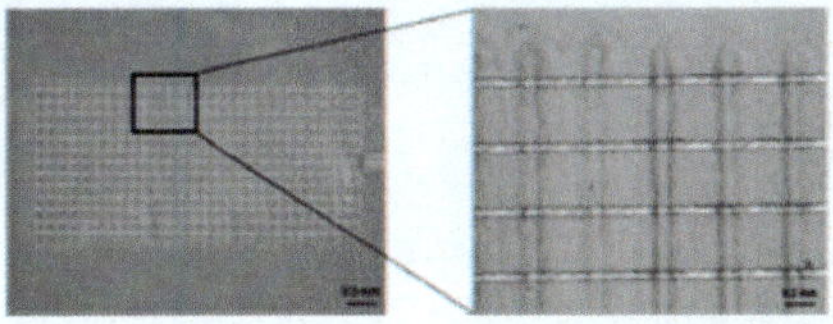

(c) AJ printed microelectrode array images at different magnifications

Fig. 17 AM printed microelectrode arrays [25]

a finite distance apart from each other as shown in Fig. 17a–c. For their developed work, Yang et al. [25] performed the comparative study and found good agreement between predicted and existing model.

The use of AM in micro-manufacturing technique has been presented by Farré-Lladós et al. [26] as shown in Fig. 18. Their developed process shows that AM can manufacture micro-channels much faster than previous micro-manufacturing techniques. Hence, this attempt takes advantage of RP feature of AM process. Their research provides a new insight to manufacture micron-sized channels using AM technique along with UV-curable glue. In micro-channels, knowledge of surface finish is highly desirable to understand the effect of flow behaviour. The main advantage of their developed micron-sized channels was that it can hold out pressures greater than 5 MPa. The main limitation of AM in micro-manufacturing is the machine resolution. Minimum cross-sections of micron-sized channels were usually based on the capacity of machine resolution. A schematic of the micro-channel and whole topography is shown in Fig. 18a–f which includes micro-channel assembly, 3D view of the micro-channel and a histogram of the different heights.

With the initiation of direction application of AM process in cold spray as shown in Fig. 19 [27], since then, use of AM technology been significantly altered and promotes the direct deposition of material in a solid state for repairing of damage component. Further, dense deposition of relatively high hardness metals such as superalloy as well as processing of oxide-free metallic parts in open air has been made much easier with the help of the direct application AM in cold spray (CS) technology [28]. Luo et al. [28] have been demonstrated mechanical properties, microstructure and deposition behaviour of cold sprayed built parts. By using their developed in-situ

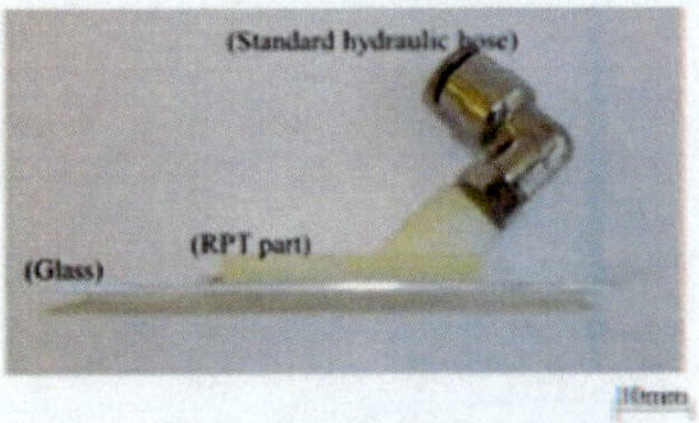

(a) View of the micro-channel assembly

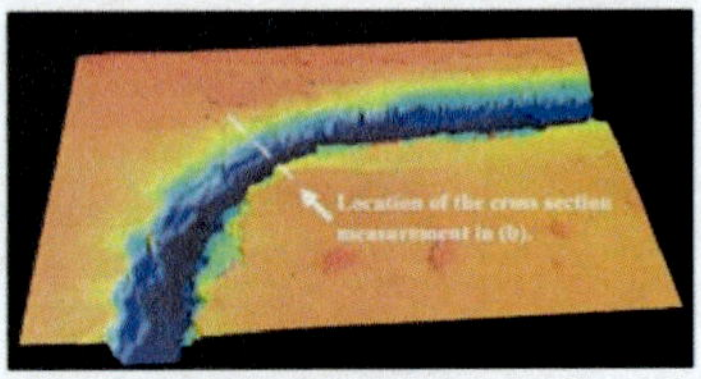

(e) 3D view of the micro-channel

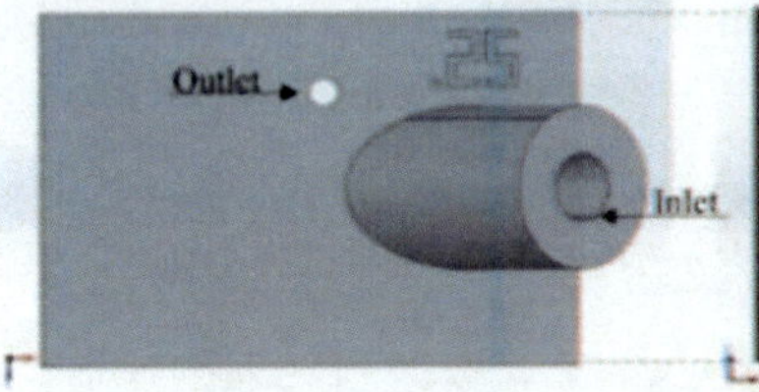

(b) RPT part top view

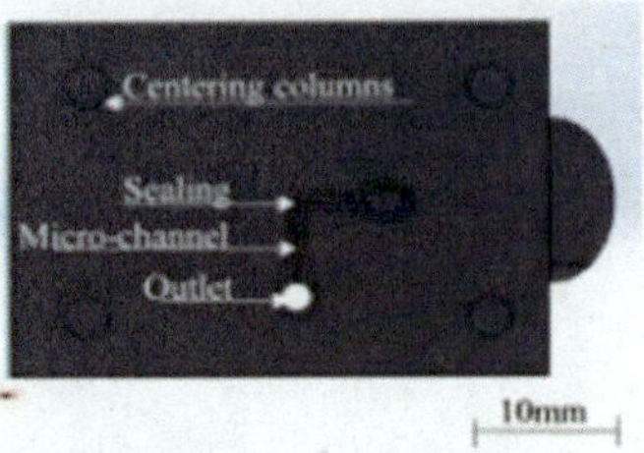

(c) RPT part bottom view

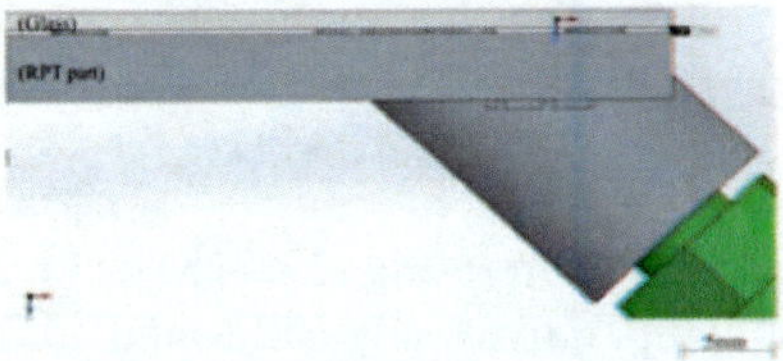

(d) Assembly detail of the gap between the RPT part and the glass

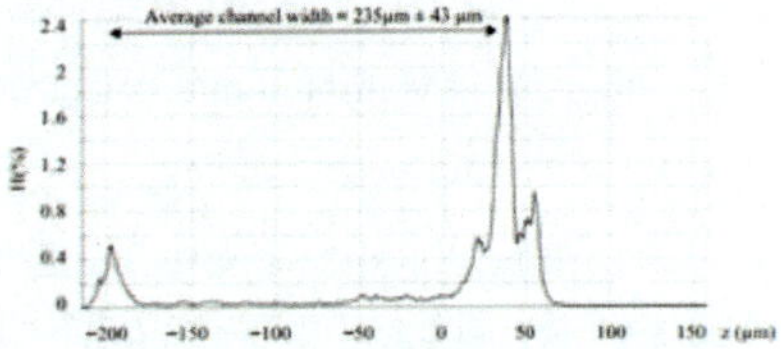

(f) Histogram of the micro-channel

Fig. 18 Use of AM in micro-manufacturing technique [26]

micro-forging-assisted CS approach they are able to improve porosity as compared to conventional CS by virtue of which they were able to improve mechanical properties as shown in Fig. 20.

Klein et al. [29] have demonstrated the application of AM in the layer-by-layer fabrication of optically transparent glass as shown in Fig. 21. They have suggested that glass printing offered a completely new application sector for AM as it can provide customised control of light and optical properties.

Borges et al. [30] have presented the use of AM technology for the production of biofuel, and it brings one of the novel and good solutions in heterogeneous reusable catalyst productions along with high mechanical strength, product separation as well as used as a basis to scale up a continuous operation in batch mode. It shows suitable alternative to homogeneous catalysts-based traditional biodiesel production process and shows high activity for the biodiesel production reaction. In their research work, they utilised AM technology for the fabrication of micro structured catalytic material which resembles the stirring system shape as shown in Fig. 22a–f. FDM process

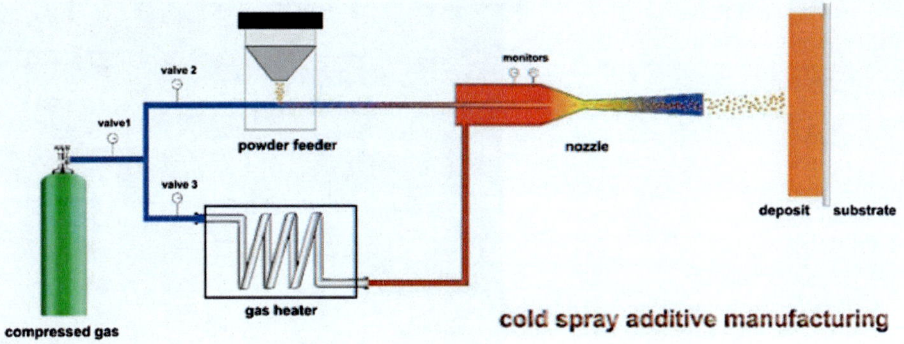

Fig. 19 Direct application of additive manufacturing concept in cold spray

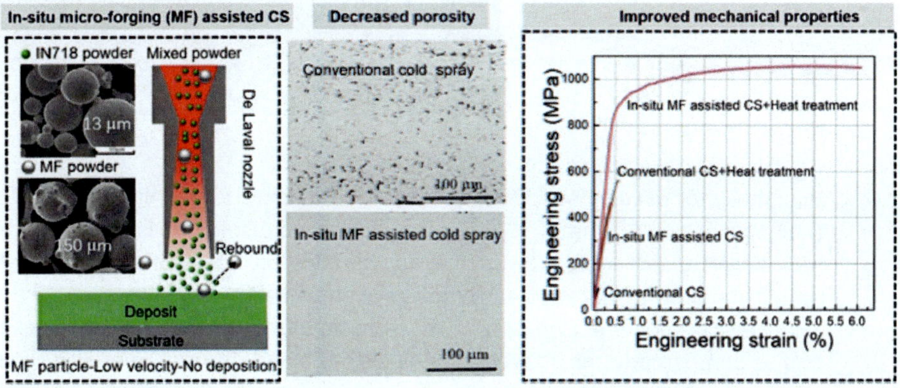

Fig. 20 In-situ micro-forging-assisted cold spray enabled additively manufactured

has been used to complete negative moulds 3D-fabrication process. With the developed system rate, differences in the reactions were found non-significant as well as it controls the mechanical damage of the catalyst significantly during continuous stirring operation induced through slurry reactor. Depending on the percentage of catalyst quantity associated with oil reactive/methanol, the yield of biodiesel or fatty acid methyl ester (FAME) found from the transesterification response was improved as shown in Fig. 22g.

Middelkoop et al. [31] have demonstrated an auspicious alternative which shows a great adsorption capacity and firm kinetics for adsorbents which were fabricated through AM process and prepared to do assembly or gas sweetening units. They have conducted packed bed reactor experiments along with 3D-printed monolith structures and commercially available 13X beads. AM fabricated monolithic sorbents along with breakthrough curves are shown in Fig. 23a. In their experiments, they have fabricated monolith structures through AM process and then loaded onto a thermocouple which is further place into the tubular reactor as shown in Fig. 23b–d. They have found desorption, and adsorption rates of monoliths are significantly faster

Fig. 21 LED-assisted caustic patterns

than those for beads. For 13X zeolite beads, they have found maximum H_2S adsorption capacity of 3.5 mmol g^{-1}. Further, they have also demonstrated that under the measured situations breakthrough tests validate that AM fabricated 13X-SU three-dimensional networks show fairly lower cyclic working capacity for H_2S than 13X-SU beads as shown in Fig. 23e. Thus, from their experimental study they suggested that as compared to the conventional adsorption and desorption rates of beads; AM fabricated three-dimensional network structures were found faster. Adsorbent beds fabricated through AM process have highly defined 3D networks which provides pressure swing adsorption and temperature swing and hence tolerating for improved mass as well as lower pressure drop and efficiency for heat transfer.

Rajasegar et al. [32] have utilized direct metal laser sintering (DMLS) for additive manufacturing of mesoscale burner array. Through DMLS they fabricated mesoscale burner by fusing GP1 stainless steel metal powder layer-by-layer technique along with 40 μm thick slice thickness. Figure 24a shows DMLS printed mesoscale burner array ready along with burner housing which is direct utilised for supply of air–fuel mixture. To demonstrate the effectiveness of the direct part fabrication through additive manufacturing process over the traditional process is also compared through Fig. 24b and its shows the importance of part fabrication through AM process. Figure 24c shows the DMLS fabricated three different designs of mesoscale burner array and shows their operation under a premixed configuration of methane (CH4) air mixture that shows uniform flames which is compact and well distributed due to the fabrication through AM process.

AM/3D printing of food is being developed by controlling the amount of printing material and nutrition content [33]. Different AM/3D food printing technologies,

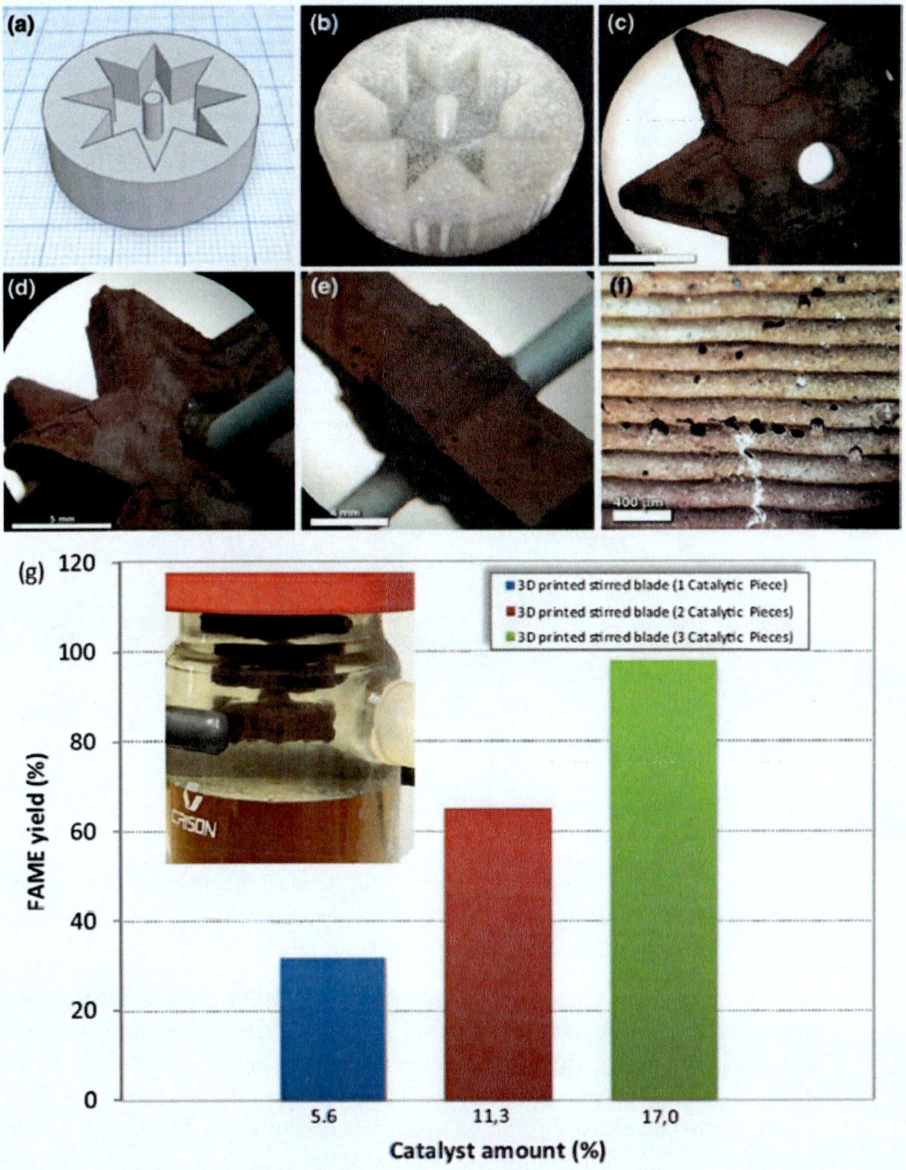

Fig. 22 AM fabricated microstructured catalytic material **a** mould design, **b** AM fabricated mould, **c–f** microstructured material **g** biodiesel yields

namely selective laser sintering/hot air sintering, hot–melt extrusion/room temperature extrusion, binder jetting and inkjet printing, were developed and shown in Fig. 25a–f, respectively. Three types of printing materials, namely natively printable materials, nonprintable traditional food materials and alternative ingredients, usually used for customised food fabrication purpose. Also, two types of recipes, namely

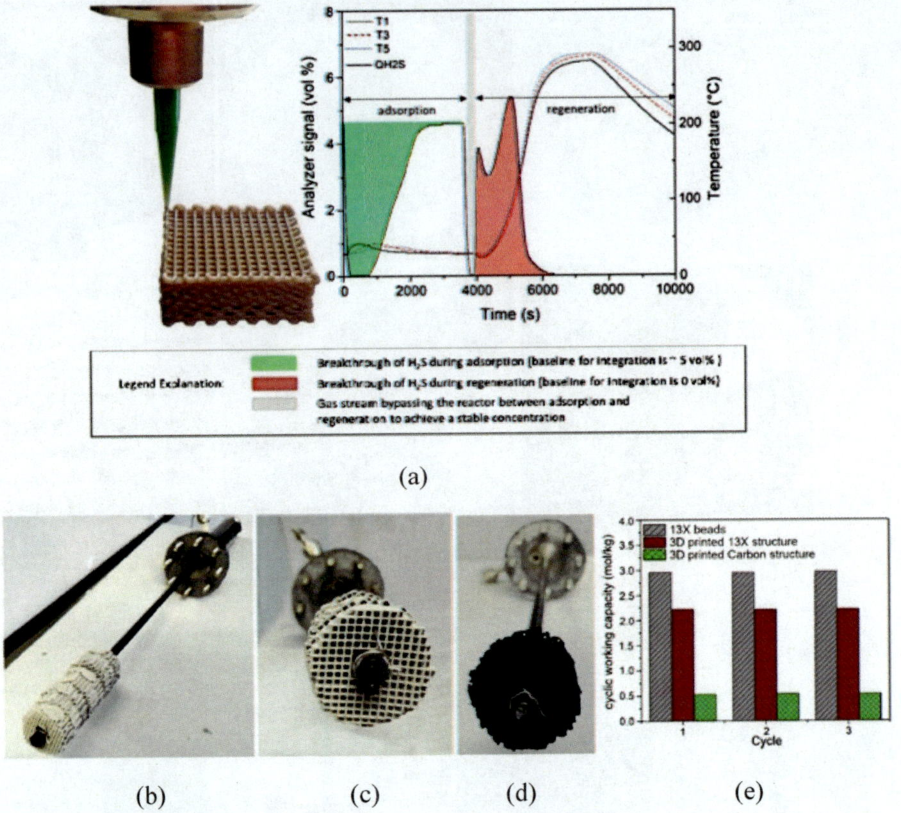

Fig. 23 AM fabricated: **a** monolithic sorbents along with their breakthrough curves, **b**, **c** 13X-SU **d** carbon monolith structures placed onto thermocouple and **e** Different sorbent materials for cyclic working capacities for H_2S

element-based recipe and traditional recipe, are often used. AM provides experimental freedom with a large variety of foods, textures and shapes where creativity is endless. The possibility of customised food printing also presented the feasibility of printing food in space. It can be used for mass production. However, food surface texture or surface finish produced by AM is one of the major issues because foods that can easily be filed are not suitable for 3D printing.

As a sociocultural system can be affected by the natural resources. Therefore, application of AM machine was also explored by researchers for sustainable development. Gwamuri et al. [34] developed a high-efficiency solar-powered 3D printer for sustainable development. In this work, they developed a system which consists of a solar photovoltaic (PV) stand-alone power/battery charging system integrated to an MOST-delta 3D printer as shown in Fig. 26a. They suggested that developed system has a great potential to control the depletion of natural resources and to increase employment, also providing a new opportunity for isolated communities by ensuring a continuous supply of rare products.

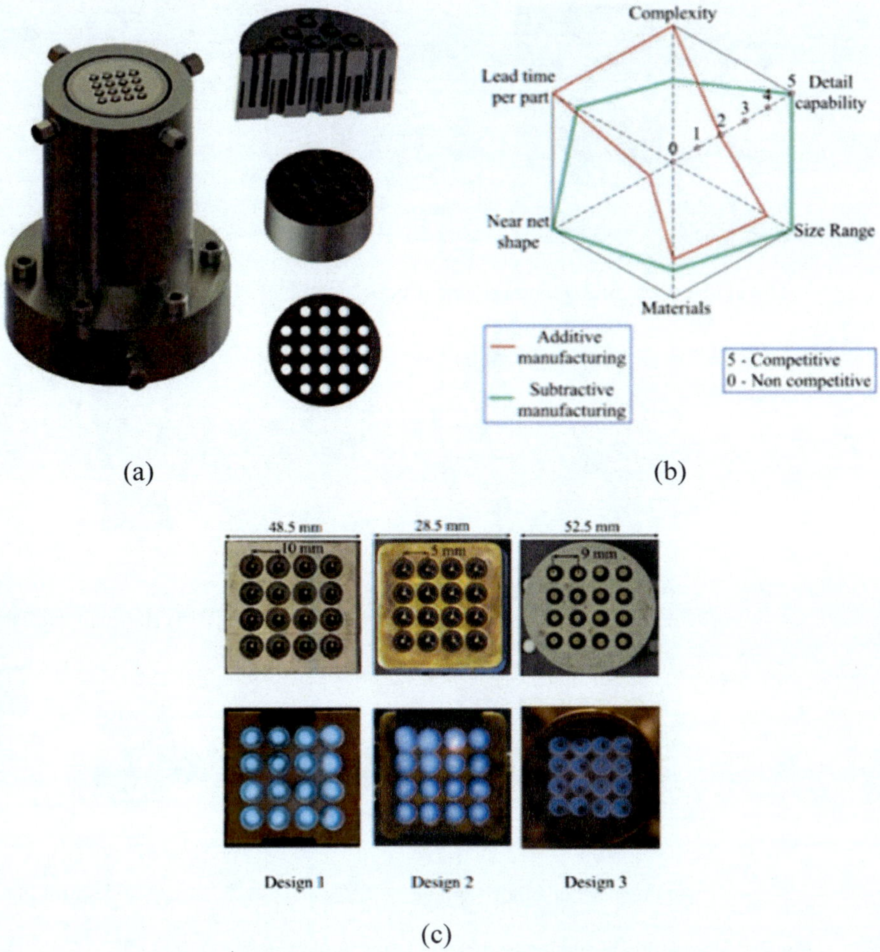

Fig. 24 **a** DMLS printed mesoscale burner array, **b** effectiveness of additive manufacturing over the traditional process and **c** DMLS fabricated three different designs of mesoscale burner array and its operation

Breddermann et al. [35] demonstrated the possibility to build pressure housings for underwater applications through the AM process. In their work, titanium hemisphere manufactured from Ti6Al4V powder through electron beam melting (EBM) process was presented as shown in Fig. 26b. From an ocean engineer's point of view, it was demonstrated that fabrication of housings for underwater applications could be explored with AM process. With AM different possibilities can be explored such as tailoring material properties [36], extraordinary geometry fabrication. Hence, AM provides a new picture of manufacturing in future [37–40]. Besides the above applications, there are some breakthrough and iconic applications of AM. Atala et al.

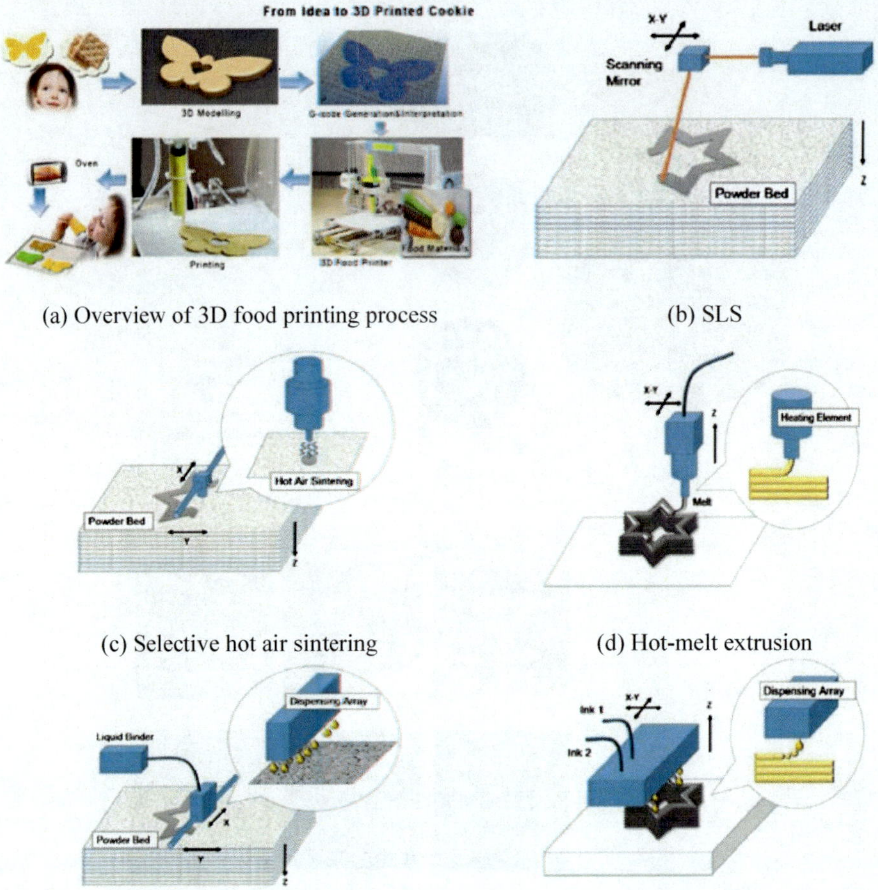

(a) Overview of 3D food printing process (b) SLS

(c) Selective hot air sintering (d) Hot-melt extrusion

(e) Powder bed binder jetting (f) Inkjet printing

Fig. 25 3D printing technologies for food fabrication [33]

[41] of wake forest institute of regenerative medicine have pioneered the use of bio-printing organs. He has developed a process for printing human kidney which is still in experimental phase. General Electric has used laser sintering process to print fuel nozzles for their LEAP® engines. They have also produced parts (please refer Fig. 26c) of GEnxt engine using direct metal laser sintering process. Also, they are using AM extensively in R and D activities. Researchers team at Monash University [42] had successfully build world's first 3D-printed small jet engine (please refer Fig. 26d). It was a breakthrough in the aerospace industry.

Some typical applications of AM in the fabrication of electromechanical devices are also presented in Fig. 26e–h. Moreover, from the above-presented studies, it has also been observed that the part surface quality can affect the different AM applications such as prototyping, rapid tooling and direct parts.

(a) Solar-powered 3D printer [33]

(b) Pressure housing [34]

(c) GEnxt engine [36]

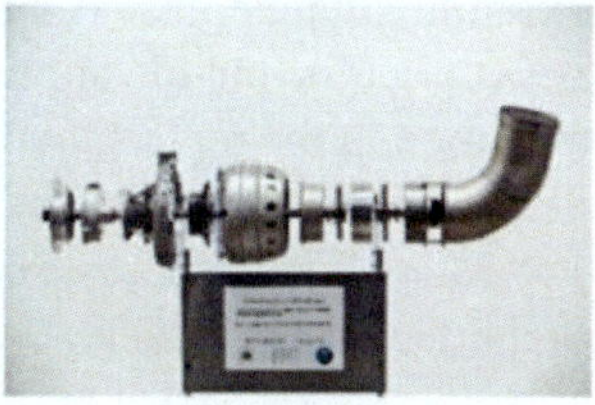

(d) 3D printed small jet engine [35]

(e) Printed Pneuflex actuator [37]

(f) Bitblox printed Infrared remote controller [38]

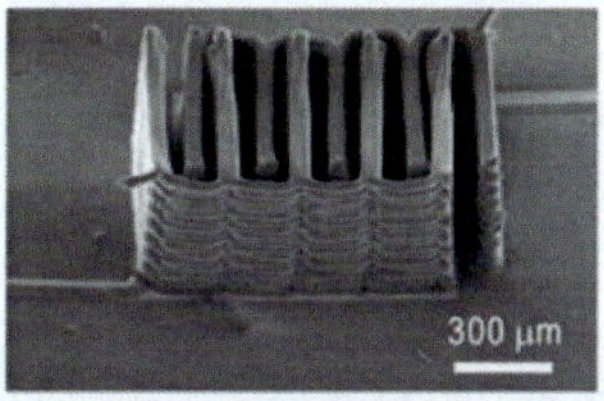

(g) 3D Printed Li-Ion battery [39]

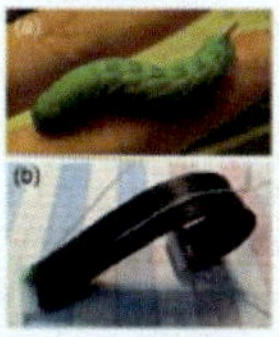

(h) Soft robot built by 3D printing [40]

Fig. 26 List of distinguished AM/3D printing applications

3 Conclusions

In the present work, a review of interdisciplinary types of AM applications has been summarised, followed by a detail explanation and critical observations of each research work. The AM applications have been classified based on their interdisciplinary works and use in prototyping, rapid tooling and direct part production rather than on the type of AM process used, although this has also not been totally neglected. The classification was made in a way that it can easily correlate the application and use of additive manufacturing in different sectors in the current scenario.

Subsequently, application and use of AM processes in the past and future trends were presented and explained, based not only on the prototyping but also on the rapid tooling and direct part production. As the most frequent application of AM process is prototyping, followed by the rapid tooling and direct part production. In past, most authors deal with prototyping and rapid tooling; however, in the current scenario trend has been changed nowadays quite a few others deal with and direct part production. The most challenging thing with AM to explore its application and use in new sectors such as adsorption and adsorbents, biofuel, mesoscale burner array. Complex and quick part fabrication is usually preferred by AM process. However, maximum studies present either an application and use of AM process in similar types of sectors, but the application and uses of AM in more innovative field and its application in the replacement of some of the conventional manufacturing processes with proper break-even point reasoning are not explored, requiring further research. The AM application sectors could significantly explore by the use of this technology in manufacturing, industry and sociocultural sectors.

References

1. Quinlan HE, Hasan T, … Hart AJ (2017) Industrial and consumer uses of additive manufacturing: a discussion of capabilities, trajectories, and challenges. J Ind Ecol 21:S15–S20
2. Lipson H, Kurman M (2013) Fabricated: the new world of 3D printing. John Wiley & Sons, Indianapolis, IN, USA
3. Noorani R (2006) Rapid prototyping: principles and applications. John Wiley & Sons Incorporated
4. Smith J, Xiong W, … Liu WK (2016) Linking process, structure, property, and performance for metal-based additive manufacturing: computational approaches with experimental support. Comput Mech 57:583–610
5. Ion A, Frohnhofen J, … Baudisch P (2016) Metamaterial mechanisms. In: UIST 2016—Proceedings of the 29th annual symposium on user interface software and technology. Association for Computing Machinery, Inc, pp 529–539
6. Ion A, Wall L, … Baudisch P (2017) Digital mechanical metamaterials. In: Conference on human factors in computing systems—proceedings. Association for Computing Machinery, pp 977–988
7. Pham D, Dimov SS (2012) Rapid manufacturing: the technologies and applications of rapid prototyping and rapid tooling. Springer Science & Business Media
8. Gibson I, Kvan T, Wai Ming L (2002) Rapid prototyping for architectural models. Rapid Prototyp J 8(2):91–95
9. Ahmad A, Darmoul S, Ameen W, Abidi MH, AL-ahmari AM (2015) Rapid prototyping for assembly training and validation. IFAC-PapersOnLine 48(3):412–417
10. Kroll E, Artzi D (2011) Enhancing aerospace engineering students' learning with 3D printing wind-tunnel models. Rapid Prototyping J 17(5):393–402
11. Minetola P, Iuliano L, Bassoli E, Gatto A (2015) Impact of additive manufacturing on engineering education–evidence from Italy. Rapid Prototyping J 21(5):535–555
12. Meisel N, Gaynor A, Williams C, Guest J (2013) Multiple-material topology optimization of compliant mechanisms created via polyjet 3d printing. In: 24th annual international solid freeform fabrication symposium–an additive manufacturing conference
13. 3D Systems. https://www.3dsystems.com/learning-center/case-studies/3d-printing-saves-day-powerade-fifa-2014. Accessed 10 July 2019

14. Makepartsfast. https://www.makepartsfast.com/rapid-prototyping-a-digital-multimeter-a-case-study. Accessed 10 July 2019
15. Materialise. https://www.materialise.com/en/cases/hoyas-vision-simulator-eyegenius-3d-printed-eyecare-devices. Accessed 10 July 2019
16. Prodways. https://www.prodways.com/en/industrial_segment/dental/orthodontic-models-for-thermoforming-of-aligners/. Accessed 10 July 2019
17. Studiofathom. https://studiofathom.com/wp-content/uploads/Vacuum-Forming-White-Paper-F001-5-1-2014.pdf. Accessed 16 May 2018
18. Bastech. http://www.bastech.com/3d-printed-injection-molding-tools-bi-link/. Accessed 10 July 2019
19. Javelin. http://www.javelin-tech.com/3d-printer/making-molded-pulp-packages-in-low-quantities/. Accessed 10 July 2019
20. Matterhackers. https://www.matterhackers.com/articles/how-professionals-are-using-3d-printed-jigs-and-fixtures. Accessed 10 July 2019
21. Mun J, Yun BG, Ju J, Chang BM (2015) Indirect additive manufacturing based casting of a periodic 3D cellular metal–flow simulation of molten aluminum alloy. J Manuf Process 17:28–40
22. Petrovic V, Vicente Haro Gonzalez J, Jorda Ferrando O, Delgado Gordillo J, Ramon Blasco Puchades J, Portoles Grinan L (2011) Additive layered manufacturing: sectors of industrial application shown through case studies. Int J Prod Res 49(4):1061–1079
23. Chen RK, Jin YA, Wensman J, Shih A (2016) Additive manufacturing of custom orthoses and prostheses—A review. Addit Manuf 12:77–89
24. Khan SF, Dalgarno KW (2009) Design of customized medical implants by layered manufacturing. NC University, UK, School of Mechanical and Systems Engineering
25. Yang H, Rahman MT, Du D, Panat R, Lin Y (2016) 3-D printed adjustable microelectrode arrays for electrochemical sensing and biosensing. Sens Actuators B Chem 230:600–606
26. Farré-Lladós J, Casals-Terré J, Voltas J, Westerberg LG (2016) The use of rapid prototyping techniques (RPT) to manufacture micro channels suitable for high operation pressures and μPIV. Rapid Prototyping J 22(1):67–76
27. Yin S, Cavaliere P, … Lupoi R (2018) Cold spray additive manufacturing and repair: fundamentals and applications. Add Manuf 21:628–650
28. Luo XT, Yao ML, … Li CJ (2018) Deposition behavior, microstructure and mechanical properties of an in-situ micro-forging assisted cold spray enabled additively manufactured Inconel 718 alloy. Mater Des 155:384–395
29. Klein J, Stern M, … Oxman N (2015) Additive manufacturing of optically transparent glass. 3D Print Add Manuf 2:92–105
30. Borges ME, Hernández L, … Esparza P (2017) Use of 3D printing for biofuel production: efficient catalyst for sustainable biodiesel production from wastes. Clean Technol Environ Policy 19:2113–2127
31. Middelkoop V, Coenen K, … Gallucci F˜(2019) 3D printed versus spherical adsorbents for gas sweetening. Chem Eng J 357:309–319
32. Rajasegar R, Mitsingas CM, … Yoo J (2018) Development and characterization of additive-manufactured mesoscale combustor array. J Energy Eng 144:04018013
33. Sun J, Zhou W, Huang D, Fuh JY, Hong GS (2015) An overview of 3D printing technologies for food fabrication. Food Bioprocess Technol 8(8):1605–1615
34. Gwamuri J, Franco D, Khan KY, Gauchia L, Pearce JM (2016) High-Efficiency
35. Breddermann K, Drescher P, Polzin C, Seitz H, Paschen M (2016) Printed pressure housings for underwater applications. Ocean Eng 113:57–63
36. Jain PK, Pandey PM, Rao PVM (2010) Tailoring material properties in layered manufacturing. Mater Design 31(7):3490–3498
37. Gibson I, Rosen DW, Stucker B (2010) Additive manufacturing technologies, vol 238. Springer, New York
38. Hopkinson N, Hague R, Dickens P (eds) (2006) Rapid manufacturing: an industrial revolution for the digital age. John Wiley & Sons

39. Conner BP, Manogharan GP, Meyers KL (2015) An assessment of implementation of entry-level 3D printers from the perspective of small businesses. Rapid Prototyp J 21(5):582–597
40. Gao W, Zhang Y, Ramanujan D, Ramani K, Chen Y, Williams CB, Wang CC, Shin YC, Zhang S, Zavattieri PD (2015) The status, challenges, and future of additive manufacturing in engineering. Comp Aided Des 69:65–89
41. Atala A, Bauer SB, Soker S, Yoo JJ, Retik AB (2006) Tissue-engineered autologous bladders for patients needing cystoplasty. Lancet 367(9518):1241–1246
42. Anon (2015b) The world's first printed jet engine. http://www.scienceinpublic.com.au/media-releases/monash-avalonairshow-2015. Accessed 9 Aug 2015
43. Anon (2015a) GEnxt engine. http://www.ge.com/stories/advanced-manufacturing. Accessed 9 Aug 2015
44. Deimel R, Brock O (2013) A compliant hand based on a novel pneumatic actuator. In: 2013 IEEE international conference on robotics and automation (ICRA). IEEE, pp 2047–2053
45. MacCurdy R, McNicoll A, Lipson H (2014) Bitblox: printable digital materials for electromechanical machines. Int J Robot Res 0278364914532149
46. Sun K, Wei TS, Ahn BY, Seo JY, Dillon SJ, Lewis JA (2013) 3D Printing of interdigitated Li-ion microbattery architectures. Adv Mater 25(33):4539–4543
47. Umedachi T, Vikas V, Trimmer BA (2013) Highly deformable 3-d printed soft robot generating inching and crawling locomotions with variable friction legs. In: 2013 IEEE/RSJ international conference on intelligent robots and systems. IEEE, pp 4590–4595
48. 3Dsystems. https://www.3dsystems.com/same-day-sheet-hydroforming-using-projet-3500-series-3d-printer. Accessed 10 July 2019
49. 3Dsystems. https://www.3dsystems.com/on-demand-manufacturing/investment-casting-patterns. Accessed 10 July 2019